Piano Action Regulating covers the action regulating requirements for the **BTEC Advanced Diploma in Piano Action Regulating** taught at the **Royal National College**.

Other books by the author include ***Theory and Practice of Piano Tuning: a manual on the art, techniques and theory***,
ISBN 978-0-9554649-0-4

Piano action regulating

a reference source for students and professionals

Brian Capleton PhD

Amarilli Books
amarilli-books.co.uk

Published by Amarilli Books
Malvern, UK

amarilli-books.co.uk

First published 2007

ISBN 978-0-9554649-1-1

A CIP catalogue record for this book is available from the British Library

Contents

Acknowledgement

The author gratefully acknowledges the Royal National College for the resources used in the preparation of material for this book.

Some equivalent terminology

Some of the terminology used in this book appears on the left. Some equivalents or an explanation is given to the right.

Blow distance - *height of rise - striking height - strike distance - hammer stroke*
Depth of touch - *key dip*
Drop (drop distance) - *hammer drop* - the distance of hammer nose from the string, when the roller is supported by the repetition lever, limited by the drop screw in the grand hammer butt.
Hammer rest felt - *support cushion*
Jack - *fly*
Lever - *whippen - support*
Repetition lever - *balancier*
Roller - *knuckle*
Set-off - *let off* - the distance of the hammer nose from the string on first escapement.

In the following list, some commonly used terms appear on the left, and the equivalent used in this book is on the right.

Balancier - repetition lever
Fly - jack
Hammer stroke - blow distance
Height of rise - blow distance
Key dip - depth of touch
Knuckle - roller
Let off - set-off (see above)
Strike distance - blow distance
Striking height - blow distance
Support - lever
Support cushion - hammer rest felt *or* lever block felt *or* heel
Whippen - lever

Upright underdamper action

1. hammer felt
2. hammer nose
3. hammer underfelt
4. hammer head
5. hammer shank
6. hammer rest baize
7. half-blow rail
8. hammer rest rail
9. half-blow rail flange
10. butt
11. balance hammer shank
12. balance hammer
13. check felt
14. check head
15. tie tape
16. tape end
17. bridle wire
18. check wire
19. lever
20. key
21. key lead
22. back touch rail
23. back touch baize
24. capstan (pilot screw)
25. heel of action box cloth
26. jack spring (spiral spring)
27. jack tail
28. set-off button
29. set-off button rail
30. set-off regulating screw
31. left & right screw
32. jack slap rail
33. notch (and jack top)
34. notch cushion
35. hammer flange
36. loop
37. butt return spring
38. damper spring
39. damper body
40. damper tail
41. spoon
42. lever flange
43. beam rail
44. damper flange
45. damper wire
46. damper stop rail
47. damper drum screw
48. damper drum
49. damper head
50. damper felt

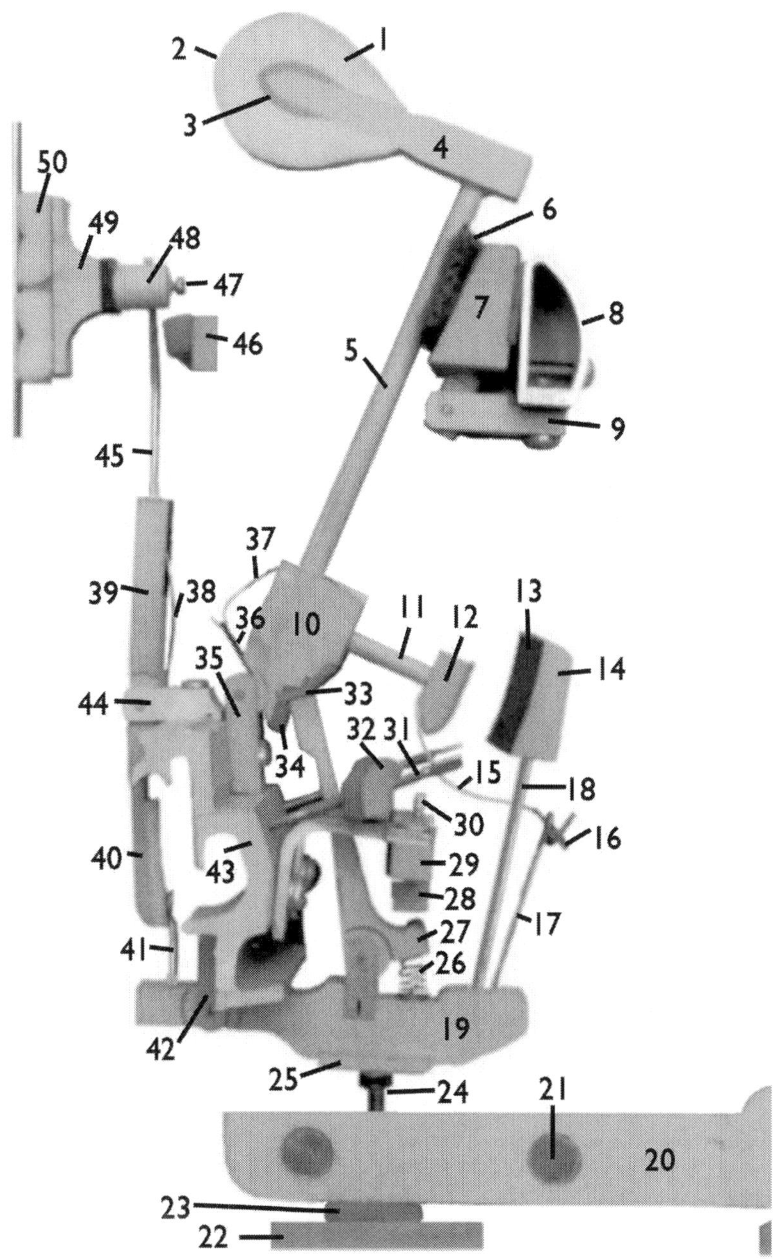
1
2
3
4
5
6
7
8
9
10
11
12
13
14
15
16
17
18
19
20
21
22
23
24
25
26
27
28
29
30
31
32
33
34
35
36
37
38
39
40
41
42
43
44
45
46
47
48
49
50

Grand roller (Érard) action

1. damper head
2. damper felt
3. damper guide rail
4. damper wire
5. hammer felt
6. hammer underfelt
7. hammer nose
8. hammer shank
9. roller
10. hammer centre
11. drop screw
12. hammer butt screw
13. hammer butt
14. butt rail
15. set-off dolly
16. jack tail
17. key chasing
18. balance pin
19. balance rail
20. balance washer
21. key
22. back touch baize
23. damper lift (spoon) felt
24. damper lift rail
25. collective damper lift regulating screw
26. collective damper lift baize
27. damper lever flange
28. damper lever
29. damper lever stop rail
30. standard damper lift
31. back check wire
32. back check leather
33. bottom lever spring and loop
34. spoon
35. capstan (pilot screw)
36. lever block
37. jack
38. jack regulating button
39. jack regulating screw
40. bent block
41. bottom lever
42. balance block
43. repetition spring
44. repetition lever
45. repetition spring regulating screw
46. hammer rest rail
47. hammer head
48. back check head
49. repetition lever regulating screw

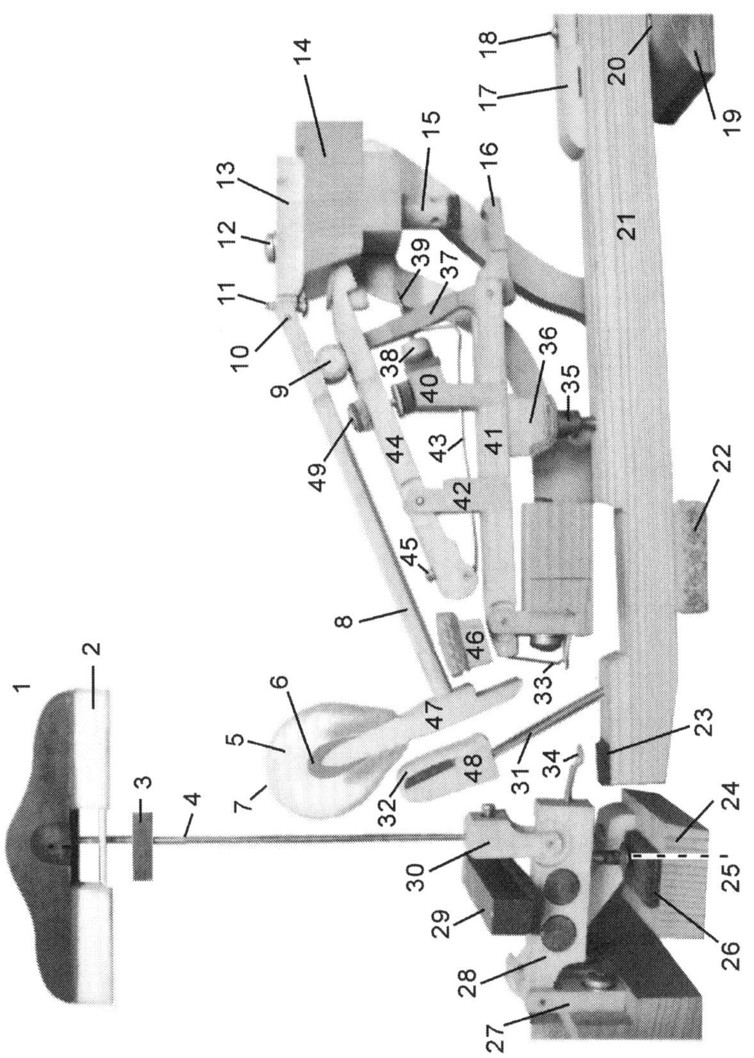
1
2
3
4
5
6
7
8
9
10
11
12
13
14
15
16
17
18
19
20
21
22
23
24
25'
26
27
28
29
30
31
32
33
34
35
36
37
38
39
40
41
42
43
44
45
46
47
48
49

Blüthner-type action (1)

1. carriage
2. abstract
3. nose
4. abstract pin
5. jack spring
6. abstract jack-top spring (jack slap spring with jack slap button)
7. set-off regulating screw
8. hammer rest rail
9. back check

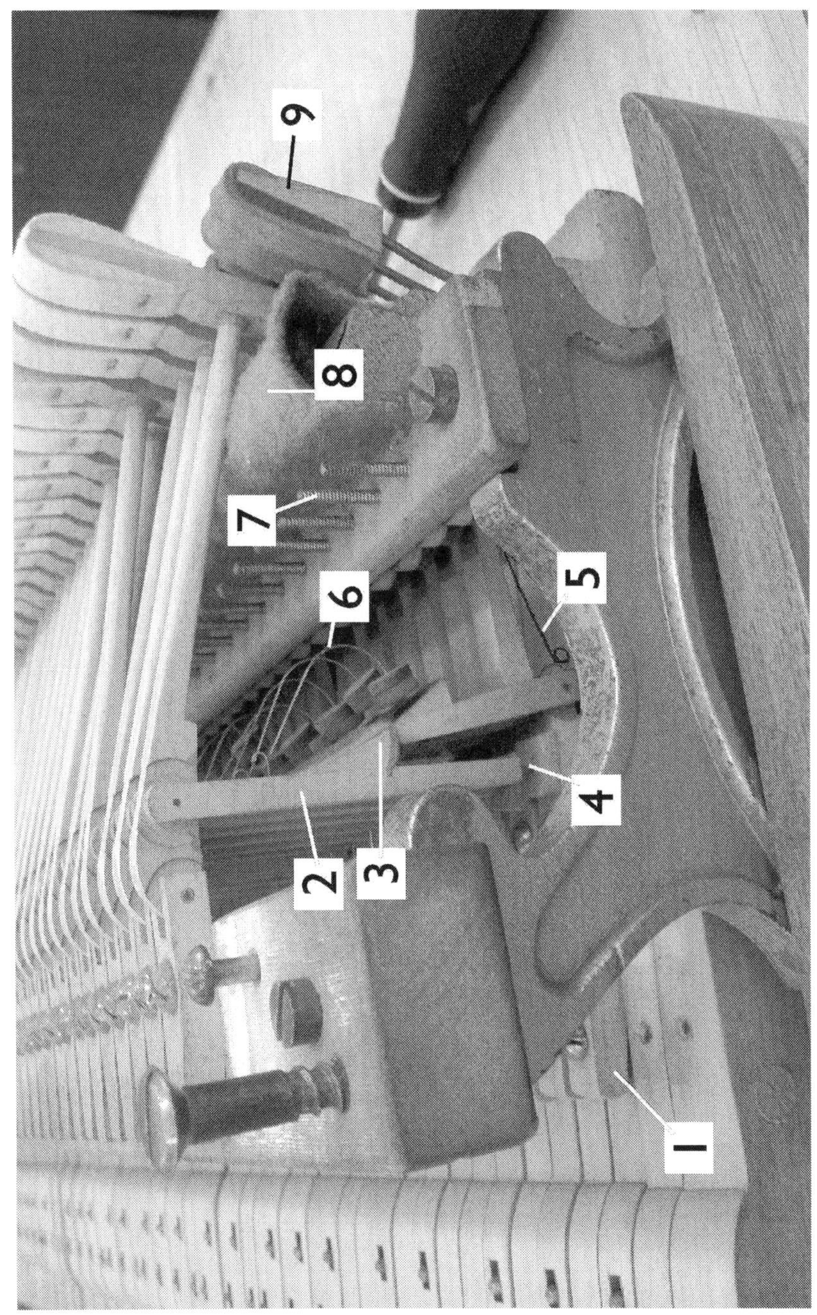
1
2
3
4
5
6
7
8
9

Blüthner-type action (2)

1. carriage
2. abstract pin
3. jack tail
4. abstract
5. set-off regulating screw
6. repetition spring

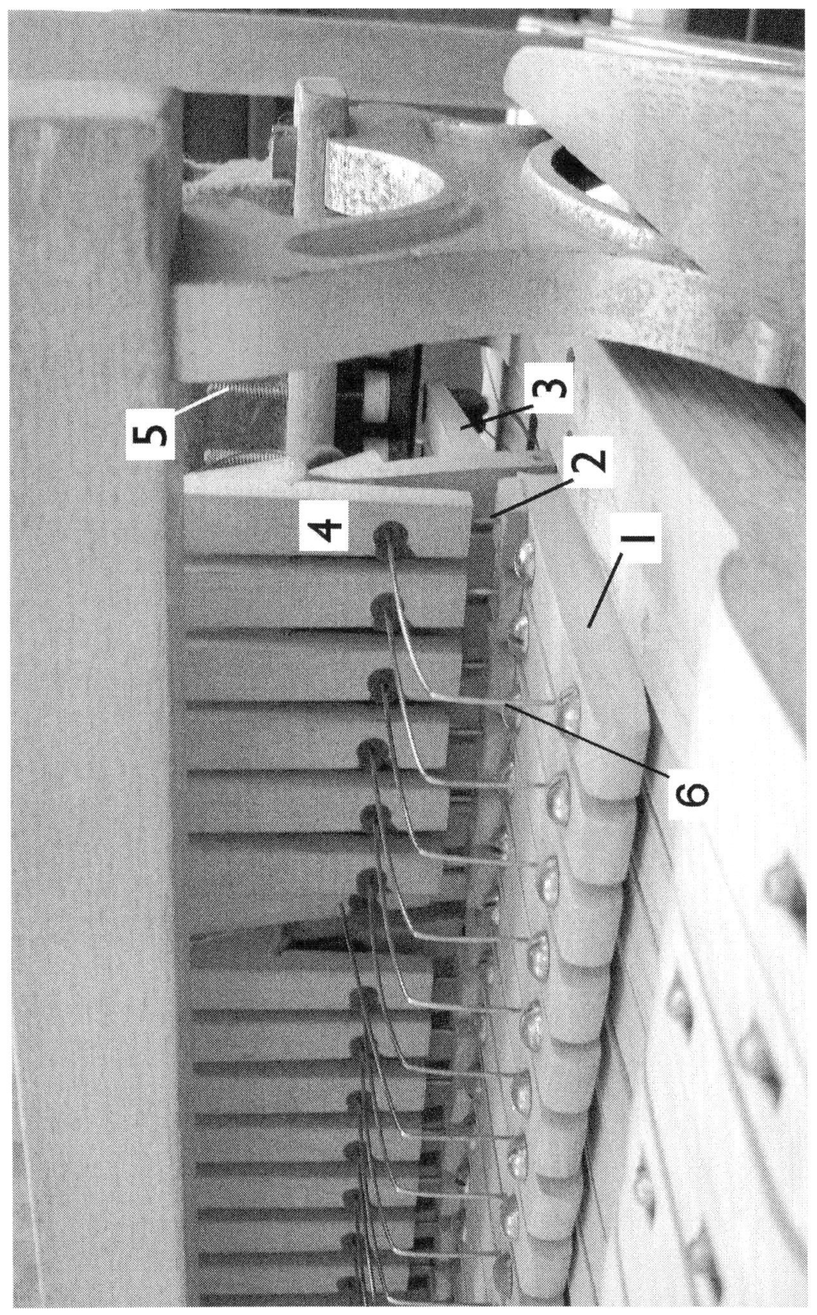
5
4
3
2
1
6

D-type action
(Simplex or spring and loop action)

1. repetition spring regulating screw
2. jack
3. loop
4. repetition spring
5. set-off regulating screw
6. hammer rest rail
7. back check

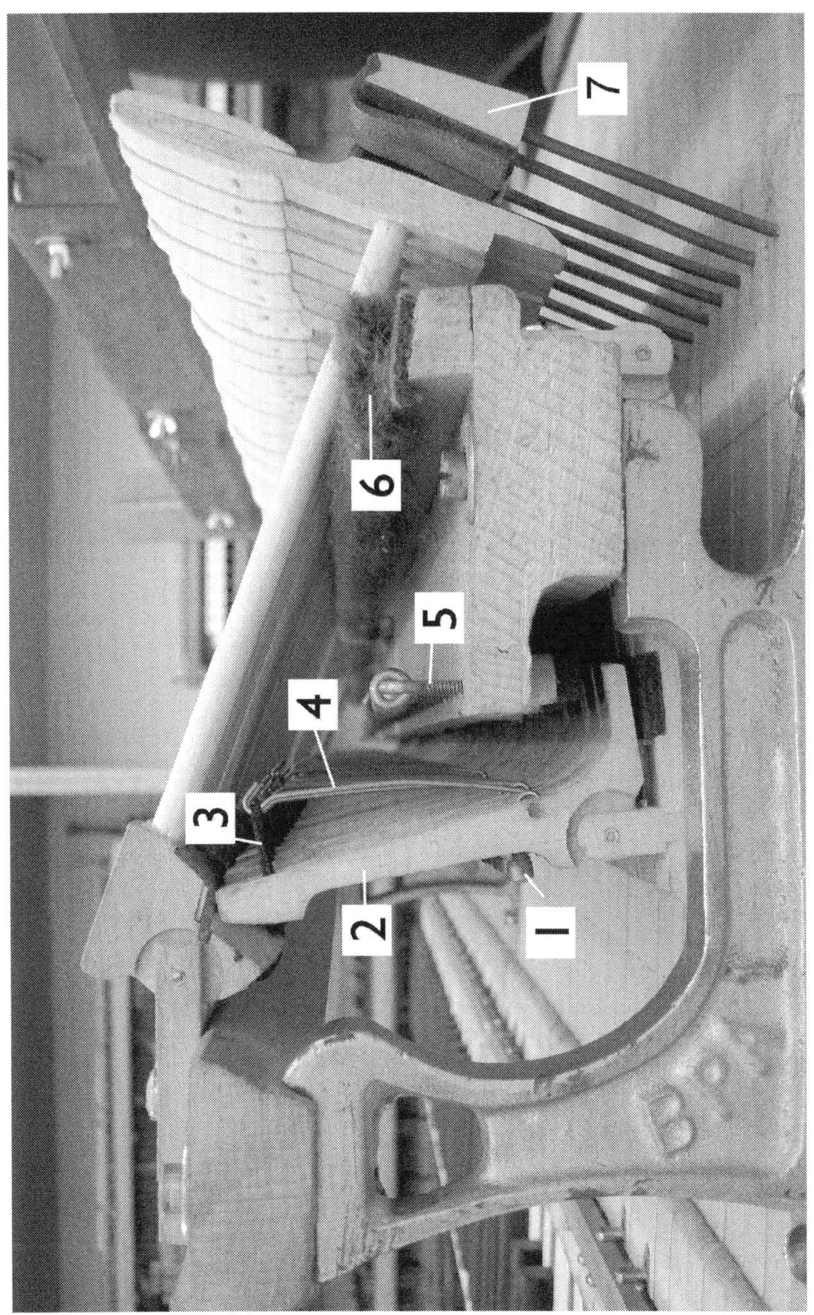
1
2
3
4
5
6
7

Standard roller action views

1. drop screw
2. set-off dolly
3. jack tail
4. balance pin
5. key keeper

Roller action with hammers lifted, for inspecting the rollers and the position of the jack top in the gate.

Standard roller action viewed from the keyboard.

The Steinway roller action

1. jack regulating button
2. repetition spring
3. repetition lever regulating button
4. hammer rest felt

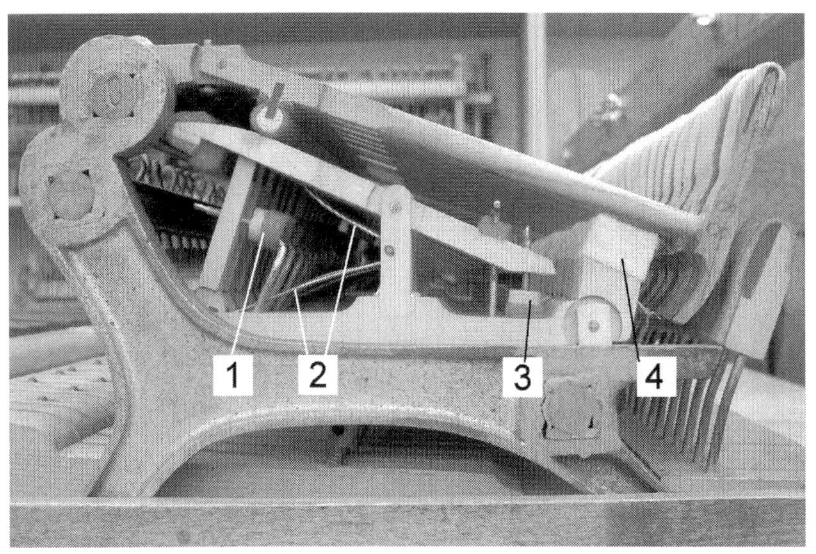

Steinway roller action, treble end view.

Steinway roller action, showing individual hammer rest felts.

Steinway roller action, showing the set-off rail.

Early Steinway *sostenuto* system: fixed lips on damper lifts.

1 - Introduction

An overview

In normal playing the pianist has no direct contact with the strings of the piano, but must rely on the piano action to covey motion from the hands to the hammers that strike the strings. The action is expected to be capable of transmitting the musical intentions that the pianist endeavours to realise through playing technique. It should therefore not be thought of merely as a machine that sits between the player and the strings, whose parts can be simply adjusted according to a set of measurements. Rather, for the finest results it is imperative that the regulation of the action is carried out bearing in mind the musical purpose of the action.

Distance measurements in the set-up or regulation of an action are design features that should be correct in the first instance. However, distances alone, set-up to the manufacturer's defaults, will not necessarily ensure the optimum operation of the action, especially if the action is not newly manufactured or in perfect condition. The reason is simply that the correct working of the action depends on both distances and the *relative timings* of the action parts that move. The action consists of a set of both stationary and moving parts, constructed of wood, sometimes plastic or nylon, cloth, felt, and other relatively "low technology" materials, that operate as a whole. The motion of the moving parts depends both on their relative positions and distances, and on other parameters including mass (or weight), friction and spring tension.

The aim is for the response and feel (the *touch*) of the action to be right, when it is used to produce music. Regularity of the basic measurements is essential in the first instance, but good regulating goes beyond this. Good regulation demands that the technician appreciates how the action should *feel* and *respond* in musical usage.

These facts mean that there are two types of assessment that the technician has to carry out when regulating. These are (1) technical tests and measurements, and (2) *pianistic tests*. Technical tests and measurements alone will not always guarantee optimum response of the action for pianistic tests.

The piano action is not a machine in which one thing happens after another, only one thing happening at a time. The operation is more complex than this, a number of things generally happening at once. We need to understand collectively what is supposed to happen in the action, mechanically, and *why*, in terms of its functional purpose as part of a musical instrument.

Musical demands

We can list the main elements of the required response of a note in the action in the following five points:

1. The note must be capable of playing a complete range of musical dynamics from very soft to very loud.
2. The note must be capable of continually repeating, for both soft and loud strikes.
3. The note must be capable of continual repetition, with the key being raised for each repetition, only a minimum distance above its fully depressed position.
4. The hammer must never "block" against the string, damping it, and it must never strike the string more than once, as a result of only a single strike by the pianist.
5. After playing a note, the action should allow the note to be freely sustained by the pianist, and should provide damping when the key has risen again to an appropriate height. Damping should then be efficient and free from excess noise.

Merely setting up an action according to the default set of recommended measurements, will not necessarily ensure the optimum meeting of all these demands for musical performance. Many of the distance and timing regulations require finer adjustments based on the actual response and feel of each individual note in the individual action.

There are a number of factors in the underlying physics of both upright and grand actions, affecting the feel and response of the action, that need to be appreciated. To begin with, every action, in operation, involves moving parts, gravity, and springs. The corresponding physical properties we are dealing with, are *friction*, spring *tension* (elastic force) and *weight* (or alternatively *mass*). In discussing action *touch* it is often convenient to talk about *inertia*, which is the tendency of the action parts to resist the attempt to move them, not due to frictional or spring forces, but specifically due to the *mass* of the action components, particularly the hammers. Scientifically, the intuitive idea of inertia would be replaced by the more rigorous concepts of *mass*, *force* and *acceleration*. As technicians, we need to be able to *feel* the distinct effects of friction, weight, inertia and spring tension in the action.

Friction

Friction can occur when two surfaces are in contact with each other. It arises as a force "resisting" relative movement, or attempted movement between the two surfaces. The two surfaces do not even have to be in actual motion against each other for friction to occur. If a force is applied that attempts to move one object against the surface of another, a type of friction called *static friction* occurs before any motion starts, resisting the start of motion.

Once motion takes place, a second kind of friction called *kinetic friction* takes over, resisting motion, or making it more difficult to maintain. If you try to push a heavy wooden block across the surface of a table, by applying a gentle push to the bottom of the block, parallel to the table surface, you may not be able to make the block move. Once you are pushing hard enough, the block will probably move suddenly, a small distance. What happens is that you first have to apply sufficient pushing force to overcome the *static friction* between the block and the table. Once the block starts moving, of you let go, it will not just slide across the table and fall off the table, because *kinetic friction* will cause it to slow down and stop. If we were to lubricate the surface of the table with a suitable liquid, say oil, then we would remove most of this friction and the block would be much more likely to end up on the floor. In piano actions, some critical friction points are partially lubricated with black lead, wax, or Teflon, to reduce the friction. These include the jack top, jack tail, the gate of the repetition lever in roller grand actions, and the top of some capstan dollies.

Generally, static friction is greater than kinetic friction. This means that at the point in time where motion starts to happen, the force moving an object against friction, is already greater than it needs to be in order to overcome kinetic friction and maintain motion. If once the motion starts, there is a tendency for the force to reduce again, then this often leads to a "stick and slip" situation. Very rapid "stick and slip" motion can sometimes occur in a piano action between the top of the jack and the roller or notch leather. This may happen if the roller or notch leather is worn, or if the lubricant on the top of the jack is worn or ineffective, and the key is being depressed very slowly. This can be *felt* in the touch, as a "creaking". This can also occur between upright damper springs and the slot in the damper body where they push on the damper.

In the piano action friction occurs in the centres, in the key bushings, between the jack top and the roller or notch, and in many other places. Being able to *feel* this friction in the touch is an important diagnostic technique. Stiff centres can make the touch feel "heavier" because of the excess friction they produce. This is particularly noticeable as a distinct resistive "feel" to the action motion when the note is played by lifting the lever rather than depressing the key. Excess friction in the key bushings can be detected by isolating the key from the rest of the action parts for that note, which is also achieved by lifting the lever off the back end of the key.

Some friction between the top of the jack and the roller or notch is to be expected, and provides part of the correct touch characteristics. It is the effects of the friction that one initially feels as the set-off. Too much friction can accentuate any "stick and slip" sensation in set-off, but it does not follow that the *elimination* of the feel of normal frictions in the action operation, would necessarily be an ideal, even if this were possible.

Inertia

Imagine now again the block on the table, and that the table surface is indeed lubricated. Let us say there is negligible friction, either static or dynamic, between the block and the table. Does this mean that a mere fly bumping into the block will cause it to shoot across the table and onto the floor? Of course not. If the block is heavy, that is, if it is made from a dense material like beech or steel, we may still need to apply a considerable force to get it moving quickly. This, however, is not because of friction, but because of the block's *inertia*. Inertia is a convenient (only semi-scientific) concept to describe the impedance that an object presents to the attempt to accelerate it, due to its *mass*, rather than to any other force like, for example, friction.

A massive object will have more inertia than one that is not massive. We tend to use the word massive in everyday language to imply "very large", but really, there is no direct implication of physical size in the strict meaning of the word, as it is used in mechanics. It is perfectly possible for a small object to be more massive than larger one. This just means the small object has more *mass*, more *quantity of matter* contained in its small size, than the larger one. Mass can be thought of as the amount of matter contained in an object, rather than implying its physical size. A small object can contain more matter than a large one, if it is more *dense* than the large one. In fact, the *density* of a material is described scientifically as its mass divided by its volume. For example, the density of hammer felt might be quoted in kilograms per cubic metre. For a given hammer volume, denser felt will result in a hammer with more mass, and hence more inertia.

Weight, mass and inertia

Pianists sometimes confuse an action's *inertia* with *friction* or even *spring tension*. "Inertia" is sometimes used incorrectly as a word to imply a negative feature of the touch. If the friction in the bushings is too great, or other frictions such as those at set-off are too great, then this is a negative effect resulting in inefficiency and more effort in playing, but it does *not* produce extra *inertia* in the touch, and the use of the word "inertia" to describe such faults is erroneous. Similarly, actual *inertia* in the touch is not in itself a negative feature, unless it is caused by incorrect key weighting (lead weights inserted in the key). We need to understand inertia in the touch, and why it is important.

It is relatively easy to show using the equations of mechanics, the importance of sufficient hammer mass. However, here we are going to examine the effects of hammer mass, and particularly the (semi-scientific) idea of *inertia* in the touch, without resorting to mathematics, as we are primarily interested in understanding these ideas from a practical point of view, rather than as action designers.

In a gravitational field, i.e. on Earth, it is *mass* that provides *weight.* Mass is not, however, the same thing as weight. The weight of an object can change in different gravitational fields, even though its mass stays the same. If we transported a piano to the moon, it would weigh a great deal less on the moon than it did on Earth, but its mass would remain the same. The piano would not have changed, but the moon's gravity is less than the Earth's gravity. Weight is the force exerted on mass, by gravity, so it depends on the strength of the gravity.

Even where there is no gravity, or where there is gravity but *gravity plays no part in the motion*, the mass of an object still remains, and to a person attempting to move the object, it would still offer *impedance to the attempt to move the object*, which is experienced as its *inertia.* This is what happen with the block on the table. It is not the gravity, or the *weight* of the block that prevents the fly bumping into it, from moving it. It is its *mass*, or in piano touch terms, its *inertia.*

In addition to mass creating inertia, *mass* also at the same time creates *weight*, so heavier objects do have more *inertia* than a light objects. Heavier hammers do therefore produce more *inertia* in the touch. The greater the inertia, the more *force* we have to apply to achieve a given hammer velocity over a given distance. The piano action with greater *inertia* (due mostly to greater hammer mass), requires more force on the keys, to achieve a given hammer velocity, than one with less inertia. This might look as though inertia is a bad thing, because it requires more playing effort, or more hand and finger "weight" to be applied to the keys by the pianist, to achieve a given hammer velocity. However, all is not what it might seem at first sight.

When the hammer strikes the string, it imparts *energy* to the string, causing the string's vibratory motion, and the subsequent sound to radiate from the soundboard. The string gets its energy from the hammer striking it. Where does the hammer get its energy from? It comes from the pianist. This energy supplied by the pianist in depressing the key must be carried by the hammer to the string as the *kinetic energy* of the hammer.

The hammer must be capable of carrying sufficient *kinetic energy* to produce the musical power and tone of which the piano is capable. The amount of kinetic energy the hammer has when it strikes the string depends on its *mass* (which is felt as inertia in the touch), and its *velocity.*

The pianist may be able to supply plenty of energy, but it can only be delivered to the strings through these two things, the hammer mass (which is felt in the touch as inertia), and the hammer velocity, which is always going to be about five times whatever speed the pianist depresses the key. The product (their multiplication together) of these two things, in mechanics, is the hammer's *momentum*, and the *kinetic energy* that the hammer carries at any time, is directly proportional to its momentum.

The result is that the smaller the *inertia* of the action, the faster the pianist has to depress the key to deliver a given quantity of energy to the strings. Conversely, the greater the inertia, the slower the key needs to be depressed to deliver the same quantity of energy. It is still *easy* to depress the key slowly, where there is greater inertia, and in fact, greater inertia enables this to be done with *greater musical control* given the right technique. Action inertia is therefore equated with the action's ability to produce more musical power without compromising its ability to produce *pianissimo* sound. In other words, inertia provides *dynamic range*, and efficient transfer of energy from the hands to the strings.

Musical power is not the only advantage of inertia. The *tone* produced when the strings are struck is dependent on the properties of the hammer head, including its mass. Too little mass will produce an inferior tone, and whilst other qualities of the felt are also of great importance, high quality felt is usually denser felt, and *inertia* can therefore be related to tone quality. The hammer head with greater mass will also behave differently in relation to the repetition process. On the grand piano it returns from the string more effectively and can generally be checked by the check head more readily.

Spring tension

An object such as a damper or a hammer on a grand piano action, with no springs attached, and without any sticky friction points in its travel, will exert the same downwards force (due to gravity), no matter where it is in its travel. Springs, on the other hand, work in a different way. The more a spring or sprung part is displaced, the greater the force it applies in the opposite direction to the direction it is being moved. Small displacements are therefore met with small returning forces, whilst large displacements are met with large returning forces. Springs therefore contribute "spongy" or inconstant touch attributes.

The most obvious place where this is encountered is in the damping mechanism of the upright piano action. The further the damper is lifted off the string, the greater the return force it exerts through the action and onto the key itself, which can be felt in the finger. The stronger the spring, the greater the effect. On most upright piano actions, the point at which the damper begins to lift can be felt in the key, and the change in spring tension from the beginning of lift to the completion of the lift, can be felt. The damper spring is a major factor in determining what is popularly (but inaccurately) called the "weight of touch".

In contrast, on the grand action with unsprung dampers (some grand actions have damper springs), the point at which the damper begins to lift may be felt through the key, but the return force exerted by the damper after this point is constant. The force is due to the weight of the damper, which remains constant no matter how far the damper is lifted.

"Weight of touch"

Most musicians will refer to the "weight of touch" of the action as an indication of how "heavy" or "light" the action feels. The subjective feeling of "heaviness" or "lightness" of the action is of course a product of many different factors in the action, that not only include actual *weights* of action components, but many other factors grouping into the four properties outlined above - *friction*, *inertia*, *weight* and *spring tension*. These properties are quite different from each other, with the result that as piano technicians we must understand that when musicians use the terms "heavy" or "light", etc., these are generic terms, and can be misleading.

It is certainly not the case that "heavy" or "light" actions, as they are so described, are in themselves necessarily good or bad, or working properly or working improperly. In particular, it is not necessarily the case that an action described as "light" is easier to play, in the sense that it is easier to produce good music, expression, or sensitive playing. It is not necessarily the case that an action described as "heavier" than another, will encourage coarser playing. The term "weight" used by musicians in the phrase "weight of touch", means many different things in different circumstances, and as technicians we need to be more specific.

The *touch* itself does not have just one characteristic of "weight", but has components of *weight*, *inertia*, *spring tension* and *friction*, and these can vary depending on how far the key is depressed, how forcefully the note is being played, and what stage in the repetition process the touch is being considered.

Of these, it is unnecessary *friction* and *spring tension*, beyond that necessary for optimum operation of the action, that destroys the *efficacy* of the action, and is to be avoided. Perceived "weight of touch" can be increased by friction and spring tensions being too great, whereupon less of the effort made by the pianist is translated into kinetic energy of the hammer, string vibrations, and finally sound. In other words, effort supplied by the pianist is wasted by the action whose perceived "weight" contains abnormally large components of friction or spring tension. Springs simply "absorb" energy (they convert the supplied kinetic energy into potential energy) and then release it again without transferring it to the strings, and friction just drains energy away by converting it into heat. The good action, conversely, passes energy from the pianist's hands most efficiently to the strings, keeping friction and spring tensions to the minimum necessary for correct operation of the action.

The presence of weight and inertia in the action, does not cause inefficiency in the process of transferring energy from the player's fingers to the strings. It may be true that more muscular effort is required to play on the action that has more inertia, or more weight, and this may not suit some pianists, but the energy supplied by the pianist is not wasted by these touch components.

Playing and *uplift* weights

"Weight of touch" in action regulating is actually quoted as *two measurements* for each single note. Although these measurements can be a useful in the regulation of both upright and grand actions during the course of their life, it must be remembered that these measurements are simple, practical *regulating specifications* rather than a complete description of the touch characteristics of an action.

The "weight of touch" in piano action regulating is in two parts, the *playing weight* and the *uplift weight*. The *playing weight* is the *minimum* weight that must be applied to the front of the key when it is in the fully raised position, in order for the key to be depressed to the point where set-off has been completed. The *uplift weight* is the *maximum* weight that can be applied to the front of the already depressed key, that still allows the key to fully rise from its *playing weight* depressed position. Both weights are determined using metal (usually brass) weight sets.

The *uplift weight* determines how rapidly the key will rise, and so has an important bearing on the action's efficacy in repetition. Both weights must exclude any effect from the damper. The repetition (or uplift weight) cannot be allowed to be dependent on forces from the dampers, because this would mean use of the sustaining pedal could impede repetition. The dampers should therefore be lifted when measuring the weights.

When the *playing weight* depresses the key, it must overcome the force due to the weight of that note's action parts, *plus* the friction in the bushings, *plus* the friction between the top of the jack and the roller or notch leather. In contrast, when the key just lifts a brass weight measuring the *uplift* weight, it is the weight of the action parts rather than the applied *uplift weight*, that must overcome the frictions in the action.

The *playing weight* will therefore always be greater than the *uplift weight*, and their difference is a measure of the *friction* present in the action. The minimum *uplift weight* necessary for good repetition is normally considered to be around 20g, whilst the *playing weight* is normally around 50g, but may be more. Measuring the *playing weight* and *uplift weight* on compared notes can determine irregularities between them, in their friction components.

Learning to detect the touch characteristics of an action through "feel", using the fingers, is at least as important as determining the playing and uplift weights, for most regulating other than in manufacturing or rebuilding. Unless one is constructing the action, fitting hammers, weighting the keys, or carrying out other initial set-up processes in manufacture or rebuilding, then the regulating weights have limited application. One does not have to be a concert pianist to understand and appreciate the requirements for a piano action's touch characteristics. Experience and knowledge of these characteristics, in the context of pianistic demands, is itself a fast and reliable method for determining what is necessary.

Changing the "weight of touch"

Excessive friction can be reduced by re-centering the flanges or changing the rollers (knuckles) in a grand action, but weighting the keys in the front or removing hammer felt is usually to be avoided. Either approach will reduce the uplift weight which will decrease the action's efficacy at note repetition. Reducing hammer mass in a grand action can result in a loss of dynamic range, and a loss of tone quality in *piano* playing.

Playing weight can be *increased* by fitting heavier hammers. Adding key weights to the *back* of the key will increase the *uplift weight* but will not increase the action's sensitivity or improve its performance, unless the keys were initially incorrectly weighted. Too much weight in the keys will increase the basic inertial feel of the touch, without any of the advantages of hammer inertia, and it can produce negative effects on the response of the action in playing.

On both upright and grand actions, the *regulation of the dampers* can have a marked effect on perceived "weight of touch", and this will be is discussed in more detail in the following chapters.

Touch characteristic

The playing and uplift weights do not fully encapsulate all the touch characteristics of an action. In particular, the playing weight is a *minimum* weight measurement for depressing the key. In normal playing, much greater forces will be applied to the key, during which the *inertia* of the action becomes an important factor determining the touch.

Both upright and grand actions share *four* essential touch characteristics. The grand action has a further two touch characteristics, making a total of six, for the grand. Although these characteristics are important, taking regulating weight measurements will not give much information about them individually. These four essential touch characteristics are detected at different, distinct stages of touch, one at a time, in progressively depressing the key slowly.

First, is the *initial "weight"* necessary to depress the key as far as the point of the beginning of damper lift. This is determined mostly by the weight and inertia of the hammer, and the friction in the bushings.

Secondly, at the *damper lift*, the perceived touch "weight" increases, as the weight of the grand damper is felt, or the damper spring tension on the upright is felt. One should be aware of the point of damper lift in relation to the depth the key has been depressed.

Thirdly, the *set-off* process offers an increasing sensation of *friction* to the touch, as the jack attempts to slide away from the roller or notch leather. This frictional feel reaches a distinct maximum and then rapidly diminishes as the jack trips away from the roller or notch leather. One should be able to feel the friction characteristics of the set-off, in "slow motion". Poor roller or notch condition, or lack of lubricant on the jack top, can often be felt as a rapid "stick and slip" creaking or grinding when the set of is made to happen very slowly and sensitively. On the upright action, the tension from the damper spring also increases during set-off.

Fourth, is the *aftertouch*, felt after set-off has completed. On the upright piano, this may include further spring tension from the damper spring, sometimes extra "spring" from the squashing of the jack slap rail felt, "spring" from the checking (the check head pushes on the balance hammer, and the check wire bends elastically a little), and finally, "spring" from the front touch baize and washers when the key is pressed fully bedded. On the grand roller action, aftertouch may include increased spring tension from the repetition spring, "spring" from the back checking as the check wire bends elastically whilst the check head in jammed on the hammer tail, and spring" from the front touch baize and washers when the key is pressed fully bedded.

In the grand roller action, a further two touch characteristics are felt in the *second escapement process*. Firstly, as the depressed key is released, one should feel for a knocking sensation in the key, that is caused by the lifting of the hammer by the repetition spring tension, as the checking is released. The knocking is caused primarily by the repetition lever's upward movement being stopped by the drop screw. The greater the tension in the repetition spring, the faster the repetition lever is travelling when it is stopped by the drop screw, and the harder the knocking.

Secondly, the set-off as a second key depression is made from the partly raised position, offers a second set-off friction touch characteristic.

Inertia in the touch

An action presents *inertia* in its touch. In playing a note, this *inertia* of the action is only felt in the touch *as the key is actually being depressed*, i.e. when it is *in motion*. It is the *mass* of the action, and in particular the *mass* of the *hammer*, that provides this inertia. Inertia in the action is not a negative feature. It provides musical "power" for *forte* playing, and it enables controlled soft playing by providing mechanical "steadying", which makes it easier to play with "even touch" (avoiding accidental "random" louder and softer notes).

Inertia comes especially from the mass of the hammer. Where the hammer is pushed up at the roller or notch, the jack is at a considerable mechanical disadvantage. This allows the hammer to move much faster than the jack, but it also means that the downwards force on the top of the jack due to the weight of the hammer, is always greater than the weight of the hammer itself. For example, lifting a grand hammer by applying upwards force directly underneath the tail of the hammer, requires much less upwards force than lifting it by applying the force to the roller. The "inertia" of the hammer experienced when playing the note, is correspondingly greater than if the jack pushed directly up on the hammer head.

There are various other mechanical advantages and disadvantages in the action, but the overall mechanical ratio between the front of the key and the hammer it ultimately moves, is determined by the ratio of the total distance the hammer moves from rest to striking the string (the *blow distance* or *strike distance*), to the total distance the key front depresses downwards (the *depth of touch* or *key dip*). This is a ratio of around 5:1. The result is that a change in hammer weight, for example, can have an effect on the touch weight around five times greater than the change to the hammer itself.

The force exerted by the jack onto the roller of a grand hammer, will cause the hammer to accelerate towards the strings. The upwards force not only needs to overcome the downwards force caused by the weight of the hammer, and the friction in the action, but also needs to *accelerate* the *mass* of the hammer upwards. For a given force applied, this acceleration will be inversely proportional to the *mass* of the hammer (from Newton's Second Law). This relationship is felt as the *inertia* in the touch, as distinct from the "playing weight", which is felt if the key is partly depressed, but held static.

Inertia in the hammer (*i.e.* the hammer mass) has several positive consequences. The heavy (more massive) hammer will be more effective in traversing the set-off distance and producing good tone, in very quiet playing. More energy is required from the pianist to make the hammer move fast for *forte* playing, but then the hammer delivers more energy to the strings, providing more musical "power" (more sound volume). After bouncing off the string, the heavier hammer carries more momentum, which can assist repetition.

Spring tensions in the touch

Spring tension in the touch can produce a number of unpleasant effects in playing. Unlike *inertia*, a force arising from spring tension is not dependent on how loud or fast we try to play. Rather, it changes according to how far the key moves. On the upright action, if the damper springs are too strong, or the dampers lift too early, the spring tension will be too prominent in the touch, producing a spongy heaviness that increases the further the key is depressed beyond the point at which the damper lifts. This "spongy" or "springy" feel that is not desirable. In the grand roller, the effect of a repetition spring that is too strong is felt acutely as the key is released after playing a note. The over-tensioned spring then forces the repetition lever upwards vigorously after the checking is released, producing an unpleasant upwards knocking sensation or shock, in the touch.

Friction in the touch

Friction in the centres, if excessive, will be felt as a "weighty" resistance to movement, even in slow key depressions, and unlike *inertia*, will cause a loss of the energy supplied by the player, so that the strike is weaker than expected, for the amount of effort required. Tone production in playing, sensitive *piano* playing, and repetition, all become more difficult. At the point of set-off, friction plays a *positive* role in the touch, provided it is not excessive. The "feel" of the set-off as a small "notchy" resistance at the set-off point, is considered a positive touch characteristic by many pianists. It becomes detrimental when there is too much friction between the jack top and roller or notch, due to worn black lead or worn notch leather or a worn roller, or if the position of the jack top is too far towards the strings, relative to the roller or notch.

Friction in the centres will *increase* the playing weight, whilst decreasing the uplift weight. When it is excessive it produces its own perceived "weight" factor in the touch, which is quite distinct from *inertia*, or *spring tension*. It can deteriorate the available repetition speed.

Symptoms of problems

Many more touch characteristics that can occur are symptoms of regulation that is incorrect or could be improved. These are more connected with the *response* characteristics of the action to certain playing situations, and these are dealt with in more detail, below, when the *pianistic tests* are applied.

2 – Universal principles

Students of piano technology often want to find a fixed "sequence" of regulations that can be infallibly applied to any action without a real need to understand the action. Indeed, many texts on regulating present regulation sequences as "step by step" instructions. In actuality, no rigid "step by step" approach will work in all cases or even necessarily produce the best result in any particular case. Whilst one can take a sequential *approach* to regulating, one must also be aware that all action parts are interconnected to, and affected by, other action parts. As a result, many of the adjustments made are *relative* adjustments rather than absolute ones, and even absolute measurements or tests can be invalidated if they are being "interfered with", by other action parts that are incorrectly positioned and not yet correctly regulated. A "step by step" sequence may work well in some circumstances, but the reason we cannot have an absolutely foolproof "step by step" sequence of regulating processes that will work in all circumstances, is simply that the action itself is not a machine that operates in a "step by step" way. It is, in fact, largely a *network* of interconnected parts, even though some things may happen "step by step" in time as the note is played.

All actions that facilitate to some degree the five musical demands outlined in the introduction, will incorporate the following principles. These include earlier action types such the various "spring and loop" upright actions, the "D-type" grand action, and the older Bluthner type grand action. The precise distances and ratios may vary with specific actions.

These following principles apply then to all modern actions, and should be absorbed and understood. They apply to actions that are assumed to be correctly "set-up", that is, the initial "regulation" and action design features that are implemented in the *manufacture process* or re-building process are already correct. If you are building or rebuilding an action and fitting it to the piano, there are more primary set-up dimensions that need to be considered first, especially the position of the strike-line, and the distance of the hammer centres from the string plane. Later in this book, we shall examine *re-positioning the strike line*, which often becomes necessary later in a piano's life, to improve the tone.

Principle 1

The blow distance is the first regulating measurement

The blow distance is the distance from the hammer nose to the string, when the hammer is at rest, before playing of the note. This is usually 45 mm - 50 mm, depending on the action and its condition, the default being 47.5 mm (1 7/8ths inch). This is the starting point for regulating a compete, already fitted action, from which everything else follows. It is possible to regulate an action with an "incorrect" blow distance, provided the regulation takes account of this. Generally, the first regulation to carry out is to obtain the correct blow distance.

Principle 2

There is a relationship between the blow distance and the depth of touch.

The depth of touch is the total distance through which the key can be depressed. The default guide for the relationship between the blow distance and the depth of touch on the modern action, is that they should be in a ratio of approximately 5:1. Thus, for a 47.5 mm blow distance, the corresponding default depth of touch would be 9.5 mm. The actual depth of touch set in fine regulating is a product of the blow distance, the ratio, *and other features of the behaviour of the action* (that we will deal with in due course).

The hammer moves approximately 5 times faster than the key, in order to strike the string with sufficient velocity. This means there is a *velocity ratio* between the hammer and the key, of approximately 5:1. Alternatively, we can say there is a mechanical advantage of the hammer over the key, of 5:1. This means that the force exerted on the front of the key is 5 times greater than the force exerted on the hammer head by the hammer shank. Alternatively, any change to the mass of the hammer head produces a proportionately larger change to the touch felt at the front of the key.

On most actions, when the key has depressed completely to the depth of touch, *aftertouch* has been produced, which is the distance the key depresses *after* set-off has been completed. The Blüthner-type action is usually the exception to this rule.

Principle 3

There must be escapement.

The hammer must *escape* from the rest of the action a short distance before it strikes the string, so that it does not block on the string as the key is still being depressed. After escapement, it must be allowed to bounce off the string unhindered. This is achieved through *set-off*, in which the top of the jack is tripped away from the notch or roller (knuckle) so that it is no longer pushing the hammer.

Principle 4

There must be checking.

Under a normal strike, the hammer must be *checked* or "caught" by the action after striking the string and should not be allowed to bounce back onto the string. On the modern upright action this is achieved by the balance hammer coming into contact with the check head. On the grand action, it is achieved by the hammer head coming into contact with the back check head. Associated with escapement and checking is *aftertouch*. This is the distance through which the key continues to be depressed after the beginning of escapement. From this point to the point at which checking takes place on a normal blow, is the *aftertouch.* There must be sufficient *depth of touch* to allow *aftertouch.*

Principle 5

There must be a repetition facility.

After a note has been struck, it must be possible for the jack to reposition itself under the roller (knuckle) or notch ready for a repeated strike, even if the key is not released to its full height. The more *sensitive* action, including especially the *double escapement* or *double repetition* action, is able to allow this for a minimum (very small) rise in the key.

Principle 6

There must be damping.

After a note is struck, the string must be damped when the key is raised to a suitable height. Damping must be efficient and "silent" in operation.

Applying the principles

Once you understand these six principles, it is usually possible to successfully regulate even an unfamiliar action type. Whatever the action, you seek to fulfil *blow distance*, *depth of touch*, *escapement*, *checking*, *repetition* and *damping*, and then make further regulations based on the *pianistic tests.*

Tolerances

In regulating we must be *painstakingly accurate*, but we must also keep the quoted "regulating distances" such as blow distance and depth of touch in perspective. The standard blow distance is often defined as 47.5 mm. We are not, however, working to the tolerance of 0.25 mm that this seems to imply! The piano action is a machine of wood, felt and cloth. It is not a precision engineered high technology engine.

The "traditional" default blow distance is actually 1 7/8ths inches, and the depth of touch 3/8ths of an inch. Some actions do have different blow distances or depths of touch specified, but these measurements are common enough to be considered the defaults. These measurements convert to the precise quoted measurements in millimetres. Thus 1 7/8ths inches converts to 47.5 mm, and 3/8 inch converts to 9.5 mm. 1 7/8ths inches is *exactly* 5 times 3/8ths of an inch, giving the velocity ratio between the hammer nose and key front as *exactly* 5:1. The actual depth of touch experienced is of course dependent on how much force is applied to keep the key depressed, and how much elasticity there is in the front touch baize and its touch washers. The blow distance may also change by more than 0.25 mm with changes in temperature and humidity.

The painstaking accuracy in regulating does not lie in distances being set to tolerances of a fraction of a millimetre. It lies in practical *equality* of measurement sets, truly *optimised* playing characteristics, and *equality* of playing characteristics from one note to the next.

The *pianistic tests*

Every piano action should "aspire", so to speak, to the same ideals of operation, but some actions attain this better than others. The definitive playing and touch characteristics, from the point of view of both the master pianist and the piano technician, are today found in the high quality grand piano roller action, in perfect condition and perfectly regulated.

We need to be aware of the *faults* to avoid. These are:

1. *Hammer blocking.* This is when the key is depressed, but when the hammer strikes the string it damps it at the same time, or damps it when the pressure already applied to the bedded key is increased.

2. *Check blocking.* This is when the hammer fails to strike the string because the checking mechanism prevents it, *i.e.* the check head comes into contact with the balance hammer (on the upright) or hammer head (on the grand), too soon.

3. *Bubbling.* This is when the hammer strikes the string two or more times rapidly, when it should only strike once.

4. *Misfiring.* This usually applies only to grand roller actions, but *can* apply to other actions under certain circumstances. Misfiring is

the condition in which usually, the note plays under a normal strike, but on an abrupt, very hard strike, it does not play. It is caused by the jack tripping away from the notch or roller too soon. Under more extreme conditions, the note may not play at all.

Any of these faults may be found very easily, just by playing the note normally. However, lack of optimum regulation may leave any of these faults as *potential*. Therefore, if the faults are not already obvious, we must still test for their potential presence. In the *pianistic tests* of an action's regulation, we *deliberately* try to *induce* either blocking or bubbling, or misfiring, by playing or repeating the note in a certain way. We are essentially carrying out some of the same playing actions that will be used in normal playing, but we are taking the demands on the piano action to the extreme.

The test for simple blocking

1. Test the note for a soft/medium strike, but keep the key bedded hard. If the hammer blocks against the string, this is a *set-off* block (set-off is too close to the string) or a *checking block* (checking is too close, or the depth of touch is too great).
2. Test the following for quiet, medium and loud notes: Play the note, keep the key bedded, and then apply very heavy finger weight to the key. If the note plays, but then the hammer touches the string, this is a *checking block* (checking is too close or the depth of touch is too great).

The repetition test – repetition blocks

3. It is not sufficient to test for good repetition just by playing fast repeated notes. An action can often facilitate fast repetition played in one way (one often sees this done using more than one finger), and appear, superficially, to be well regulated, when in fact it will still allow repetition blocks if a different playing technique is used. Repetition excellence is not defined by speed alone. The "speed" test is in fact relatively crude, and unrevealing about the operation of the action. If repetition is properly honed and optimised, which can be done on relatively slow tests, then

the facility for repetition at maximum speed will naturally follow. Test the following for quiet, medium and loud notes: Play the note, keep the key bedded, raise it slowly about half way, and then rapidly depress the key again from the half-raised position. Experiment with the speed at which the key is allowed to rise before the subsequent hard depression from the partly raised position. If the note does not block, repeat the test increasing its severity by raising less and less each time, until you find a point of key rise at which the note will not repeat. Sometimes, you may need to raise the key *more* than half way before depressing again, to find the block. Once the block is *induced* in this way, the distance the key can then be rapidly depressed and raised can be increased whilst the blocking continues unabated. This is a *secondary checking block* or *repetition block.* The hammer checks at a secondary distance, greater than the normal checking distance, preventing the hammer reaching the string. It is primarily a *timing* problem. The hammer is able to return too far from the string, before the jack attempts to return under the notch or roller. It can be caused by both incorrect regulation *distances* and incorrect *frictions* and *spring tensions.*

The tests for hammer bubbling

1. Play the note very gently, as if with a finger that is barely strong enough to play, bedding the key, but very tentatively as though afraid to squash whatever might be underneath the key.
2. Repeat the above test, but from an only partly raised key position. Experiment with different heights of rise.

In both cases we are attempting to *induce* "bubbling", in which the hammer bounces rapidly back onto the string after the first strike, causing a second or multiple gentle strikes. On upright pianos, be careful not to allow the finger to "cheat" as it depresses the key, and cause bubbling "unfairly". This is when, if you look carefully, what is actually happening is that the finger rapidly stops at the set-off point (or sometimes the damper lift point), *rises* a micro amount, and then continues to depress the key. This can happen because the sudden increase in touch "weight" at the set-off point allows the lightly pressing finger to "bounce" a little at this point.

The bubbling can be caused by incorrect regulation distances, the jack top not clearing the notch or roller on set-off, by a failure in the checking, and/or by spring tension that is too strong.

The test for misfiring

Strike the key *very hard*, and very suddenly, *starting from above* the key. Do not touch the key first, and then play. Repeat a number of times. Remember that misfiring can be intermittent - making just one or two tests alone is not sufficient.

The aims

If any pianistic test produces blocking, bubbling or misfiring, then the cause must be determined. The rules are that

1. No bubbling should be allowed.
2. No misfiring should be allowed.
3. The tendency for blocking must either be eliminated (the preferable position) or minimised to a degree that it can only be induced under severe testing conditions that the pianist is unlikely to reproduce.

Misfiring usually only applies to the grand action. Blocking, bubbling and misfiring are all related. On both the grand and upright actions, blocking and bubbling are, in most situations, *opposite* effects. The factors that work towards *avoiding* one of these effects, will usually work towards *causing* the other. Much good regulating, especially on a "problem" action, involves finding the critical region in which neither blocking nor bubbling can be induced.

Damping efficiency

Sometimes dampers can cause problems by being inefficient, even though the regulation seems to be adequate. The following principles apply to all dampers.

1. The felt must not be too hard, anywhere on the surface that touches the strings. Even a small spot of hardness due to dirt,

foreign matter or previous exposure to damp or fluids, can cause buzzing.

2. The felt must be "square" with the damper head.
3. The surface of the felt must be equidistant from the strings when the damper is lifted.
4. The damper must have sufficient *follow through*, that is, when it rests on the stings, and the strings are pushed in towards the soundboard, the damper head (not just the felt) must follow the motion through.
5. The damper must press with equal force at both ends of the damper, and both sides of the damper. It must not lift one side or end sooner than the other.
6. The damper must not rotate relative to the line of the string, as it lifts, or as it falls back into place.

If a damper seems to persist in allowing a "ringing on" after it returns into place, first confirm that the "ringing on" is indeed being generated from that string. Touch the string with a large part of the hand some distance away from the damper, after the damper has replaced, and see if the ringing stops. If it does not, the string has actually excited a partial from one or more of the other strings, often in the bass. Play the note hard and let the damper return. Use the palm of the hand to dampen off other sections of the stringing until the section containing the ringing string is found. Having located the culprit string(s), the corresponding damper(s) can be regulated.

When the "ringing on" is coming from the note whose damper is being regulated, first ascertain whether it is just one of the strings of the trichord (or bichord), or whether more than one string is involved. If a change to the regulation is made, and the problem continues, be sure to check that the problem has not just shifted from one string to another. Knowing which string is causing the problem, can determine which way to alter the regulation of the damper. Each string must be fully contacted by the felt. If the middle string of a trichord fails to be damped properly by a flat felt, the felt can be manipulated in the middle using toning needles to improve the result.

For a persistent ringing on of one partial, locate on the string the node point (or points) for that partial that lies under or close to the damper. The string needs to be damped *away* from the node point(s).

Stubborn problems near the central break on an overstrung piano may require the position of the damper along the line of the string, to be altered.

Some upright pianos have *fly dampers* fitted close to the break, which damp the string above the strike line, in addition to the underdampers. The fly damper should ideally be positioned a different distance from the strike line, to the main underdamper.

3 - The upright underdamper in motion

1. As the key front is depressed the back of the key rises. The *capstan* or *dolly* pushes upwards on the *heel of action*, at the bottom of the *lever*. As the lever rises, the top of the jack pushes upwards on the notch leather and in doing so pushes the hammer forward towards the strings.

2. For the notes with dampers, the spoon at the back of the key encounters the tail of the damper body, and pushes on it outwards towards the strings. The damper head then begins to lift away from the strings. At the point the damper starts to *lift*, the tension of the *damper spring* that normally holds the damper felt on the strings, starts to contribute to the *touch* felt in the key, making the touch feel "heavier" at this point.

3. The *tail* of the *jack* then encounters the *set-off button*. From this point onwards, the tail of the jack cannot rise any further, even though the *lever* continues to rise. The rising lever carries the *jack centre*, and as the jack centre continues to rise with the lever, and the jack tail is prevented from rising by the set-off button, the

jack is tripped backwards so that the *jack top* slides out from under the *notch leather*. This is the *set-off* process. The set-off process is necessary because without it, the hammer would continue to be pushed all the way to the strings, and would be *blocked* against the strings if the key front was held down fully depressed. As the set-off occurs, another change in the *touch* can be felt. The frictional sliding of the jack top from under the notch leather can be felt in the touch, even though the jack top is "lubricated" with *black-lead* (graphite paste or fluid) or Teflon.

4. What happens next depends on how fast the hammer is moving. After the jack has *set-off*, when the hammer nose is a few millimetres from the strings, there is no longer any transmission force from the pianist's finger to the hammer. The hammer must then travel under its own momentum to the string. If it has insufficient momentum, it should fall back until the *balance hammer* encounters the *check head*, unless the key is allowed to rise, and hence the lever to fall again. The nap in the *balance hammer leather* helps jam the check head against the balance hammer whenever the two meet. When they are jammed or held together, the action is said to be *in check*. Any playing weight on the key must be released in order to release the check condition.

5. Under normal playing, the hammer crosses the *set of distance* under its own momentum, strikes the string, and bounces off. The balance hammer then encounters the check head and the action *goes into check*.

6. At *check*, the jack is tripped back from the notch, the hammer is closer to the string than the full blow distance, and the lever is raised. From this position, releasing the key fully will replace all the action components back to their starting positions, ready for another note to be played. However, it is necessary to be able to play a repeated note *without* fully releasing the key front back to its full height. For this note repetition to be allowed the jack top must be able to get back under the notch, so that it can push the hammer forward again. As the key front is raised a little, the check head and balance hammer loose their connection. At the same time, the jack tail lowers away from the set-off button,

enabling the *jack spring* (spiral spring) to push the jack back under the notch.

7. The operation of the damper is independent from these other activities in the action. The damper simply lifts when the key is depressed by a certain distance, and returns to the string when the key is raised a certain distance. Nevertheless, an important damper regulating measurement, the *damper lift distance*, is the distance of the hammer nose from the string, when the damper starts to lift, when playing the note from the initial resting position.

Clearly, the efficacy with which the jack can get back under the notch in relation to how far up the key is raised after playing the note, depends on how fast the hammer tends to fall back away from the strings after check is released, how fast and with what force the jack moves when being pushed by its spring, what the alignment of the jack and notch is when the action is in check, and what the precise shape of the notch is. The fine regulating of the action will therefore require a slightly different approach with each action, using the pianistic tests together with the formal regulation measurements and adjustments.

4 - The grand roller action in motion

1. As the key front is depressed the capstan at the back of the key pushes up on the lever block (heel), raising the lever assembly. The lever assembly then raises the roller and the hammer is lifted towards the strings. Normally the upper surface of the repetition lever gate is set a very small distance above the top of the jack, so that for a very soft or slow motion strike the lever may push the roller up without assistance from the jack. In normal and *forte* playing the small distance between the top surface of the gate and the top of the gate is lost as the lever accelerates, so the jack pushes the roller.

2. For notes with dampers, when the key is around 1/3 to 1/2 depressed, the back of the key will encounter the damper lever or the spoon (depending on the action) and the damper will start to lift, increasing the force necessary to depress the key.

3. When the hammer is at the *drop distance* from the string the drop leather on the top of the repetition lever under the drop screw, encounters the drop screw, and the repetition lever is prevented from rising further, as the key continues to be depressed. The bottom lever is therefore pushed towards the repetition lever,

increasing the tension in the repetition spring. At the same time the jack top now rises beyond the gate, continuing to push up on the roller, and to push the hammer towards the string. The hammer at this point has still not *escaped*, as it is still being pushed by the jack.

4. The jack tail then encounters the set-off dolly, and although the bottom lever continues to rise, raising the jack, the jack tail cannot rise because it is prevented from doing so by the set-off dolly, so the jack is tripped back, and the jack top slides away from under the roller. This increases the repetition spring tension further. When the jack has slipped away from the roller and is no longer pushing on the roller, the hammer has then *escaped*. This is the first *escapement*. The hammer now travels under its own momentum to the strings, strikes the strings, and bounces off freely.

5. It is also possible for the hammer to escape from the lever by virtue of the fact that the key has not been fully depressed. When this happens, the hammer bounces of the strings, or even may fall back without a strike, and the roller encounters the repetition lever. The lever then absorbs some of the hammer impact because it is able to move towards the bottom lever, and the repetition spring tension allows some bouncing motion of the hammer to take place. On a properly regulated action the bounce is not sufficient to cause hammer bubbling against the string.

6. In normal playing, as the hammer is on its way back down from strings, the roller encounters the repetition lever which is still touching the drop screw as the key is still depressed. The downwards force from the roller now pushes the repetition lever downwards towards the bottom lever, increasing the repetition spring tension. The tail of the hammer jams against the back check head as the hammer goes into *check*. This keeps the knuckle forcing the repetition lever down against the repetition spring tension. The increased tension also increases the spring's force on the jack, in a direction towards the roller.

7. As the player releases the downwards force on the front of the key, the check head moves away from the tail of the hammer, the increased repetition spring tension is released, and the repetition

lever is allowed to rise again. This pushes the roller upwards, and the hammer towards the strings, and allows the jack back under the roller, pulled by the repetition spring. The upwards movement of the repetition lever is limited by the drop screw regulation. How far under the roller the jack top goes, depends on how far the key is lifted, but in the finest regulation, the smallest rise of the key allows the jack sufficiently far under the roller to enable a strike repetition, and a second set-off to take place. The position under the roller that the jack can attain, depends on how far the repetition lever rises, and is ultimately limited by the jack regulating screw.

8. Once the jack is sufficiently far under the roller, the hammer can be raised again, and a repeat strike made. The set-off distance (the smallest distance of the hammer nose from the string when the jack is no longer pushing the roller) remains *the same*, and the hammer will set-off at the same point, on the repetition strike. The relationship between the set-off distance, drop distance and jack position regulation, and its effect on repetition, is dealt with in more detail in *Fine regulating the roller grand action*.

5 - Regulating the upright underdamper action

Preparations

The following instructions presuppose that any necessary preliminary work such as cleaning, reshaping hammers, repairs, or tightening of screws, has already been carried out, and that the action is complete and intact. In manufacturing or reassembling after rebuilding, further set-up procedures are required, and the dampers are usually fitted and regulated before the other action parts.

Action position

Begin by ensuring the action is in the correct position and is properly fixed or bolted at all its securing points. No measurements can be relied on until this is done.

Pedals

Make sure that the pedals are not adjusted so that they are partially "on", or interfering with the action in any way. If necessary, de-regulate them so that they are not working, or even remove the pedal rods.

Hammer travel

Each hammer travels in an arc, and the arc should fall in a fixed plane at right angles to the plane of the strings. Pushing two or more hammers towards the strings, the gap between any two adjacent hammers at the back of the hammer head, and at the front, should remain constant as both hammers travel towards the strings. Hammers that move sideways as they approach the strings, need flange papers added between the hammer flange and the beam rail, on the same side they are moving towards, viewed from the piano playing position.

Lever travel

The top of the jack must be centrally under the notch. Jack tops that are to one side must have flange papers added between the lever flange and the beam rail on the same side, viewed from the piano playing position.

Hammer spacing

Each hammer must be in line with its strings when the hammer touches them. Make sure angled hammers on overstringing are positioned so that the string(s) fall symmetrically across the face of the hammer. On the bichords the hammer must strike both strings equally on the *face* of the hammer, and not just on one of its edges.

Blow distance

Before the blow distance can be measured, we must make sure that the hammers are actually resting on the hammer rest rail. If any hammers are not resting on the rest rail, investigate. The hammers may be sticking together, needing hammer spacing (see below), or the capstan or pilot may be screwed up too high. Whatever the cause, it needs to be corrected.

Measure the blow distance and correct as necessary by adding a strip or blocks of felt or cloth behind the half-blow rail. The default blow distance should be 47.5 mm but can be from 45 mm to 50 mm depending on the following other considerations. Having set-up the blow distance for one or two test notes at the default distance, it is necessary to check that it will actually work for this particular action. Choose a suitable test note that is a natural, and make sure that the tie tape is not taut. Push in the bridle wire if necessary. Now eliminate any lost motion between the capstan and the heel of action, or between the top of the jack and the notch leather. Do not turn up the capstan so much that the hammer is being pushed forward off the rest rail. Lost motion can be felt in the touch at the key front, by *very gently* pressing the key to find it. A capstan that is minimally too high can be detected by pressing down on the back of the key and seeing if the hammer moves backwards slightly. There should be no such movement in the hammer, and no feeling of lost motion in the touch.

Now check the height of the key above the key bed or key frame. There may be a specific design height for an individual instrument, measured from the key bed, but the height of the key above the lock rail is an important aesthetic feature that should not in any case be compromised. The key bottom should be 2 mm - 3 mm below the top of the lock rail. The rest of the regulation should accommodate easily whatever is the specific chosen height. Key height is adjusted by adding or subtracting paper washers at the balance pin, underneath the felt or cloth washer.

Set-up the depth of touch on the test note to the default depth of 1/5th the blow distance, by adding or subtracting paper washers underneath the front touch baize, at the front touch pin. If necessary change the thickness of the front touch baize. If a touch block is used, level the white keys each side of the test note to it, and place the touch block on the test key, depressing it. Press only lightly on the block. The top of the block should then be level or just slightly below the level of the adjacent keys. After this, re-check the lost motion and adjust if necessary.

Now we need to test to ensure that there is *aftertouch*. Adjust the set-off button so that the hammer sets off about 2 mm from the string. If the hammer blocks on the way to the string because the balance hammer encounters the check head before the hammer reaches the string, bend back the check wire. Depressing the key with normal force should allow set-off to take place, and the hammer to drop back to the check position. If this does not happen, try lifting the back of the lever to play the note, rather than depressing the key. This should verify that the set-off and checking process itself is able to take place. If the process does not complete when depressing the key, there is insufficient depth of touch, or the blow distance must be reduced. If this is the case, how much must the depth of touch be increased before sufficient aftertouch is attained? If the depth of touch needs to be, say, 12 mm, then one should definitely consider reducing the blow distance.

Key dropping and easing

Every key should "drop" without resistance over the balance pin, but should allow no vertical rotation or forward and back movement. The balance bushing can be tested for left and right movement in the key, and tightened if necessary using a balance bushing punch, or if it is too tight can be eased using key easing pliers. The "dropping" of the key over the balance pin should only be assessed when the balance pin is in good clean condition. It can be polished with metal polish if necessary. If it is necessary to ease the balance hole at the bottom of the key to achieve an easy key drop over the pin, this should be done with caution, so as not to oversize the hole.

At the front touch bushing, the key should not "knock" sideways when the key is raised or fully depressed, but the key should still be able to move up or down under only its own weight (without the weight of the action applied to the capstan) freely. This can be regulated by turning the front pin, or if necessary, easing the bushing using key easing pliers.

Lost motion

Lost motion throughout, can now be regulated as for the test notes (see *Blow distance*, above).

Key levelling

The height of the keys above the key bed will depend on the individual instrument. The lower front edge of the white keys must be below the level of the top of the lock rail throughout by 2 - 3 mm (no more than 3 mm). The visible part of the key fronts above the lock rail can be square. Within this range the precise height of the key is not mechanically critical, unlike many other measurements. The aesthetic details of level keys are very important.

The white keys can be levelled using a straight edge. Sometimes an edge that has a 1 mm arch is used, to place the keys in the central keyboard 1 mm higher. The black keys can then be levelled by setting them level with the white keys at the back of the key, or slightly higher. It is important that the bottom of the black key top (where it joins the main wooden key body) is slightly recessed below the level of the adjacent white keys.

Preliminary depth of touch

The depth of touch of all the white keys can now be given a preliminary regulation, using a touch block or measuring at the key front. Add paper touch washers *under* the front touch baize to reduce the depth of touch, or remove then to increase it. If necessary, the front touch baize thickness can be decreased or increased. The number of paper or card washers under the baize should be kept to minimum for the necessary depth of touch, by using the most appropriate thickness washers. A large number of thin washers will be much more compressible ("squashy") than a couple of thicker ones, and this is to be avoided. Later in the regulating it can limit how close the checking can be set, because it can allow blocking of the hammer on the string after checking, if more finger weight is applied to the depressed key.

The black keys can be regulated so that they remain about 2 mm above the surface of the adjacent white keys when they are fully depressed. The designed depth of touch for many actions may be 9 mm or 9.5 mm, so a preliminary depth of touch of a centimetre will suffice. If using a touch block, then when the key is depressed it should be possible to push the surface of the bock beneath the level of the adjacent white keys with a light playing eight on the finger.

Set-off (escapement)

The set-off distance is the *closest* distance of the hammer nose from the string, that the hammer reaches before falling back, on the very slow motion *set-off test*. Set is tested with a *slow motion* hammer movement towards the string, either by depressing the key or lifting the lever. The hammer motion must be *very slow*, so that the hammer falls back away from the string without reaching it. Testing set-off, one should *feel* in the key (or lever) the frictional resistance as the top of the jack slides away from under the notch. The default is 3 mm from the string, but for finer regulation, the distance may be reduced as far as 1 mm in the treble, and half the diameter of the string in the bass. The closer the set-off, the more sensitive the action will be in *pianissimo* playing, but the more liable it will be to the hammer blocking or bubbling against the string, especially if small changes to the regulation take place with changes of humidity, bedding in, or wear.

The set-off distance is increased by screwing the set-off screw clockwise, observed from above, and decreased by screwing anticlockwise. Screwing clockwise lowers the set of button, so that the jack tail encounters the button sooner as the lever rises. Anticlockwise raises the button, so the jack tail encounters the button later, when the hammer is closer to the string.

For set-off to be tested accurately and regulated, the hammer must be able to escape *and fall back* from the string when it is at the set-off distance from the string. There must be *aftertouch*, which is the ability of the hammer to fall back a definite distance into a checked position after set-off. If no aftertouch can be found even after screwing down the set-off button, then set-off should be tested by lifting the lever, rather than depressing the key, because insufficient depth of touch when depressing the key, can prevent the lever rising high enough to complete the tripping back of the jack away from the notch. If set-off still appears not to happen, or there is no aftertouch even when lifting the lever, then (1) ensure that the check head does not interfere with the balance hammer prior to the hammer striking the string, before testing the set-off, and (2) check that the jack slap rail is not preventing the jack from tripping back far enough away from the notch. If necessary (1) bend back the check wire, or (2) screw out the jack slap rail so that it is not interfering with the jack.

If there is no aftertouch because there is not sufficient depth of touch to allow set-off to complete, increase the depth of touch until there is definite aftertouch.

Checking and depth of touch

The checking distance is the distance of the hammer nose from the string after playing a medium force note, the hammer has successfully gone into check (i.e. the balance hammer is jammed against the check head), and the key is held down under normal playing weight. A checking distance that is too small allows one set of malfunctions whilst a checking distance that is too large, allows another set.

Choose a white test note and regulate the check by bending the check wire, so that the checking distance is 12 mm - 15 mm for a medium hard strike. Quick regulation can be done bending the whole check wire by pushing or pulling the check head, once the technique of handling the "spring" and plasticity in the wire has been acquired. This achieves much the same result as bending the wire with a wire regulating tool at the bottom of the check wire.

Bending the wire at the top, just underneath the check head (using a regulating tool), alters the *angle* of the check head relative to the balance hammer. This changes the *difference* between the checking distances for soft and hard strikes. Bending the wire *out* decreases the checking distance for a hard strike, *when the soft strike distance remains the same*. In other words, bending the wire out will initially alter both soft and hard checking distances. The soft checking distance can then be put back by bending the wire at the bottom. The hard checking distance will then have been reduced. Conversely, bending the check wire *in* at the top, *increases* the hard checking distance, once the soft checking distance has been put back where it began, by bending at the bottom. One should normally regulate so that the soft and hard checking distances are same.

The checking distance is related to the depth of touch, and later will be fine regulated through this. It is also related to the regulation of the jack slap rail. To begin with, the jack slap rail should be screwed out, away from the jack, so that it does not interfere with the jack at any time.

12 mm - 15 mm is the default checking distance. However, the *principles* must be understood, and occasionally a checking distance outside this range may be necessary. The principles of checking distance are as follows:

Up to a *minimum* checking distance *limit point*, the closer the hammer is to the string when it checks, the better the repetition capability of the action (the smaller the distance the key will have to be raised after a strike, before a repetition can be played). If the checking distance is reduced so much that it approaches the *set-off distance*, the correct operation of the action will be impeded. The check head will tend to encounter the balance hammer *before* the hammer strikes the string. On many actions, long before the checking distance is reduced this far, hammer bubbling against the string will occur on a slowly played soft notes, or blocking of the hammer against the string will occur on hard playing, or on soft playing followed by a hard pressure placed on the key. These three response conditions *must be tested for*, and avoided. They are the "checking distance too small" tests.

There will be a minimum checking distance necessary to avoid these "checking distance too small" malfunctions. The precise limit point on any action depends on the design and condition of the action, and varies from action to action. It is determined not just by distances but also by *timings*, i.e., how fast the action parts move relative to each other. Changes to the timing caused by humidity changes or other causes like wear, compression or wood movement, can subsequently increase the position of the limit point, so that the chosen regulation causes malfunction. It is not therefore advisable to place the checking distance *at* the limit point. In general, 12 mm works well.

If the checking distance is too large, good repetition can be lost by "blocking", caused by secondary checking (repetition blocking). We must discover if the action is going to allow secondary checking, by attempting to deliberately *induce* it in the following way: A note is struck and the hammer goes into check. If at this point the key is then not released, but extra pressure applied, the effect of secondary checking can often be then induced in a more pronounced way. Applying extra pressure forces the jack (fly) further back away from the notch. If the key is then slowly raised about half way (or less if necessary), and then depressed forcefully, the check head jams against the balance hammer when the hammer is still *on its way towards* the string - *secondary checking.* Reducing the checking distance will overcome this problem or improve the response by reducing the distance the key has to be raised in order to induce this blocking. We are looking for the *minimum* rise before blocking can be induced, or a condition in which no blocking can be induced. *However*, in between each change of check position, the *other* "checking distance too small" tests must be made. The "checking distance too small" malfunctions must *never* be allowed, even if this means that hammer blocking (secondary checking) can still be induced. It is not generally advisable to use just the "checking distance too small" tests or just the blocking test alone. The interplay between the two can be very fine.

Having satisfactorily regulated the test note (or test notes) the jack slap rail should be temporarily regulated just for the test note (or notes) in order to ensure the desired result for the jack slap rail regulation is attainable for the chosen checking distance. Refer below to *Repetition - the jack slap rail*, for this regulation.

Take another white test note some distance away, and regulate to the same checking distance as the first. Are the two check heads in line when the notes are at rest? Test the distance from the check head to the balance hammer in each case. Correct if necessary. Now re-regulate the checking distances (of the hammers from the strings when the hammers are in check) by altering the depths of touch, until they are the same. We should end up with both checking distances the same, *and* the check heads perfectly in line.

Now all the other check heads can be lined up between the two test notes, and the line extended across the entire compass, if necessary setting up more test notes in the same way. The entire set of check heads should then fall in a perfectly straight line.

After aligning the check heads, all the checking distances can be regulated by altering the depths of touch. Test by playing three adjacent notes at a time, and test each note using soft, medium and hard strikes. If necessary, reduce any differences between soft and hard strikes by altering the angle of the check heads as outlined above. Uniformity of both checking distance across the compass, and alignment of check heads, is the aim.

Collective damper lift

The damper stop rail is on many actions effectively "cosmetic" and needs no regulation, but sometimes it limits the amount of damper lift so that the dampers do not encounter the hammers. We begin here by assuming the damper stop rail has been so adjusted if necessary.

Firstly, as far as the dampers themselves are concerned, the damper felts must all be correctly aligned with the corresponding strings. Any vertical or lateral errors must be corrected. The damper drums should be parallel, horizontal, and at right angles to the strings. With the dampers lifted (it may be necessary to adjust the damper pedal first) the top and bottom of each damper should be equal distances from the strings. This is regulated by bending the wire at the top. Left and right overall positioning of the damper head should be made by bending the wire at the bottom. The lateral angle of the damper relative to the strings (it should be parallel to the strings) should be made by bending the wire at the top.

Damper felts should be soft, and any buzzing as the damper touches the vibrating string, caused by hardening of the felt, can be cured by replacing the felt or softening by pricking (quite extensively if necessary) with toning needles. Wedges or split wedges that do not fall properly into place in the bichord or trichord respectively, even when their alignment is correct, can be treated by compressing the wedge edges with ordinary pliers (or the tool designed for the purpose). Split wedges can have the split depth increased with a knife.

Ensure that the damper springs are all in place operating evenly. Weak springs can sometimes be strengthened by bending appropriately.

Now the *collective lift* can be regulated by bending at the bottom of the damper wire. First ensure that no damper spoons are bent so far out that they are touching the damper tail. Test to make sure no note is obviously heavy in touch due to the damper lifting immediately, or almost immediately that the note depression starts. If necessary, or if in doubt, bend the damper spoons in (away from the strings) first). Beware that bending the spoons in too far can cause the lever (whippen) to become partially lifted, causing lost motion (a gap between the capstan and the heel of action).

Well spaced test dampers must now be set-up so that there is sufficient *follow through*. When the string is pressed hard in towards the soundboard, the damper head (not just the felt) must follow the motion through, fully and readily. If it does not, the damper wire must be bent at the bottom so that it does. At the same time, when the pedal is operated, the dampers must lift fully clear of the strings. Too much follow through may prevent this.

The action can then be unbolted and leant away from the strings, or removed from the piano. A screwdriver can be inserted at the pedal's damper lift rod end of the action, behind the damper lift bar that runs horizontally along the action behind the damper tails. This causes all the dampers to be lifted as though the pedal were in operation. Now the red line of backing cloths at the backs of the damper felts can be used as a visual indication that the dampers are in line, or not as the case may be. Careful regulation of the damper wires must be carried out until the line is straight, between the test dampers already set-up. The action should then be returned to the piano.

Now the pedal itself should be used to test the collective lift. Tweak the pedal continually so that damper just start to lift, and look for those that lift earlier (sooner) or later than the test dampers. These need fine regulating by bending the damper wire at the bottom in towards the strings, to make them lift later, or away from the strings, to make them lift earlier. Double check decisions with the red line of backing cloths, to confirm that the move will *improve* the line. Red backing cloths can only be expected to be *out of line* if there is a discrepancy in the thickness of the damper felts, or if any damper heads have been replaced with ones of different dimensions. Finally, all dampers should begin to lift simultaneously, on use of the pedal.

Individual damper lift – damper spoons

The damper spoons on the backs of the levers (whippens) can now be regulated. The damper should not start to lift until the hammer is *at least* half way towards the string. Aim to *feel* the damper lift in the touch as well as observing it. The later the damper lifts, the lighter the touch will seem overall, whilst the earlier it lifts, the heavier the touch will seem. The default distance (from the hammer nose to the string at the beginning of damper lift) to aim for initially is about 20 mm. However, always note how far down the key is depressed at the point lifting begins.

The point at which the damper has fully repositioned itself and fully damped the string, as the key is *rising*, is equally as important, and *will not necessarily coincide with the lifting point.* After the strike, become aware of the how far the key has raised, at the points damping starts, and at the point it has fully completed. Damping completion should not be left until the key fully raised, or close to this point. However, the touch should also not be too heavy. The earlier the damper lift starts, the later it will complete when the key is raised again, and *vice versa*. It should start with the key no less than half way down, and complete when the key is about half way up. If it completes too early, the note will shut off too abruptly particularly on light or *staccato* playing. Regulation involves the touch, the point of re-damping after the strike, and the damper lift distance (from string to hammer). The damper lifts should be regulated so that all three are satisfactory, and there is no abrupt change of regulation from one note to the next.

Damping tests.

The dampers should work in concert with each other, as if as one unit. Pedalling and playing loud chords, and then releasing the pedal, should result in correspondingly efficient stopping of sound. There may be some overall *damped resonance* but there should be no “ringing on” due to non damping, or improper damping. Individual strings that are inefficiently damped can be quickly found by running a fingernail briskly over the whole damped part of the stringing compass, with the dampers applied. If a note appears to ring on slightly, try plucking the strings individually to determine which string may be causing the problem. Ensure that there is *follow through.* Check the collective lift regulation, the alignment of the head parallel to the strings, full cover of the strings, equal top and bottom alignment and distance from the strings, felt softness, and finally, damper spring tension.

Repetition - the jack slap rail

If the jack (fly) slap rail is screwed in towards the jack too far, the jack cannot escape sufficiently far back away from the notch during the set-off (escapement) process. The notch may then strike the top of the jack after the hammer bounces off the string, causing the hammer to bounce back towards the string and strike it again. Rapid recurrence of this causes "bubbling" of the hammer against the string.

If the rail is screwed too far out beyond the jack, it has no effect. On some instruments the position of the jack slap rail can obscure access to the set-off screws, and may need to be moved before regulating set-offs. Remember that the jack slap rail is usually in three or four sections, one for each break or stringing section of the compass, and each rail is typically supported by three *left and right* screws. A left and right screw has two threads, one in the beam rail and one in the jack slap rail itself, one being a *right hand thread* and the other being a *left hand thread*. This causes the rail to move further in or out than would be expected from an ordinary screw for each turn of the screw. On some instruments clockwise turning moves the rail in towards the jack, whilst on others the reverse applies.

The function of the jack slap rail is to limit the distance the jack escapes away from the notch during set-off, in order to enable better repetition. Repetition should be tested by playing the note and then repeating, allowing the key to rise each time only a minimal distance. One should learn to *induce* "blocking", or to *find* any "blocking" that the action is going to allow, by repeatedly raising the key a little, slowly, and then rapidly depressing it again, trying to make the hammer strike. As soon as a strike is missed, the key can probably then be slowly raised somewhat more, and then rapidly depressed, increasing the raising distance over which the note will repeatedly not play, but blocks (the key depresses with a spongy feel but no note is produced).

The blocking is caused by one of two things: (a) Secondary checking taking place due to the hammer *falling* back as the key is raised, rather than *bouncing* back off the string and going into normal check. The secondary checking prevents the jack (fly) from getting back underneath the notch until the key is fully released, and the lever (whippen) is allowed to fully fall. Alternatively (b) the hammer remains closer to the string, but the jack is remains too far back to provide an effective upwards force on the notch.

The easiest way to regulate the jack slap rail is therefore as follows:

1. Screw out the rail until it is not interfering with the jacks (flys) at any time.
2. Screw the rail in until bubbling just starts occurring on soft, medium or loud notes.
3. Screw out the rail until bubbling just stops.

Remember to test both loud and soft notes, at various positions along each rail. The rails are not necessarily straight, or screwed in equally at each *left and right* screw, and different notes may require different settings. Since there are only usually around three *left and right* screws per rail, some compromise may be called for.

If regulating the rail for a test note during the checking distance regulation, and a satisfactory result cannot be found, the checking distance may sometimes be revised. If, for example, there is either obvious blocking *or* definite bubbling, then the choice is either to allow the blocking (the bubbling cannot be allowed), or to revise the checking distance. It may, for example, be possible to reduce the checking distance and avoid blocking on the "checking distance too small" tests by altering the angle of the check head.

Further tweaks

The tendency to block can sometimes be further reduced by increasing the jack (fly) spring (spiral spring) tension and/or weakening the butt return spring. The butt return spring should first be disconnected from its loop in order to ascertain the effect. Both adjustments may also have a tendency to increase the chances of bubbling. Before taking such measures, it is advisable to ensure that sluggish jack timing contributing to blocking, is not being caused by stiff jack (fly) or lever (whippen) centres, that can otherwise be corrected by re-centering or applying a lubricating/cleaning fluid designed for piano actions.

The pedals

The modern upright piano has two or three pedals. The damper lift pedal (sustaining pedal) on the right should be regulated so that there is a very small amount of lost motion (free play) when the pedal is depressed, before the dampers start to lift. Some pianists may prefer more lost motion. The half-blow pedal on the left should operate as soon as the pedal is depressed, but should not be pushing the half blow rail forward when the pedal is raised.

The middle pedal is the celeste or practise pedal. When not in operation, the celeste felt must not come into contact with any hammer. When in full operation (the pedal is designed to be either "on" or "off", with no graduation) the celeste felt must not come into contact with any damper, and the celeste rail itself must not come into contact with any hammer.

Some upright pianos have a different arrangement such as the "sostenuto" pedal that is not really a sostenuto pedal operation as on a grand piano, but rather, lifts just the bass dampers. Other more unusual pedals on uprights include the una corda (more often found on overdamper pianos) and the real sostenuto. As on the grand piano, the una corda must generally be regulated so that on full operation the hammers in the treble strike two out of the three trichord strings, and one of the two bichord strings in the bass. The sostenuto is designed to be operated after chosen keys have been depressed. Depressing the pedal after depressing chosen keys will then keep the dampers on the chosen notes lifted, without affecting any other notes.

6 - Regulating the grand roller action

Regulating a grand roller (Érard) action has two stages, regulation on the regulating table, and fine regulation of the action *in situ*, in the piano. The regulating table substitutes for the key bed and strings, and allows full access around the action, for observation and adjustments.

Cautions for withdrawing the action

When withdrawing the action from a grand piano it is important to keep a visual check right across the compass by looking through the action well to the hammers, as the action is being withdrawn. Any hammer that rises even a small amount, will be likely to get hooked on the frame behind the wrest plank as the action is withdrawn. Any hammers that get caught while the action is in motion being withdrawn, will snap off or be damaged. Only when all the hammers are *under* the wrest plank is it safe not to look. Hammers can easily be raised just by touching the keys or other action parts whilst attempting to hold and pull the action. If the hammers lie slightly above the height of the wrest plank underside, then raising the front of the action can sometimes help lower the hammers relative to the wrest plank.

When the action is withdrawn, it can be placed on the regulating table. Any moving or lifting of the action, however, should be done with due care for health and safety. The grand action is both a heavy object and an awkward shape to lift or manoeuvre. Lift with two persons, if necessary.

Setting up the regulating table

The string height above the key bed must be measured at the beginning and end of each stringing section, and the measurements transferred to the strike bar on the regulating table.

Screw tightening

Tighten all flange and butt screws as necessary.

Key frame slides

If the key frame is not very evenly in contact with the regulating table surface, the slides should be adjusted to take up any free play. Later these will probably then need re-adjusting when the action is on the piano key bed. Testing for gaps beneath the slides and the key bed or table can be carried out by knocking the key frame with a hammer and listening for rattling.

Key easing and dropping

If a full key levelling, or other keyboard regulations need to be carried out, then the action must be removed from the keyboard, and the key keeper removed. The keys can be removed and the key frame cleaned as necessary. New balance and/or front touch baizes can be fitted if necessary. The balance pins and front touch pins must be firm, clean and smooth. If not they must be cleaned, polished, or replaced.

Key dropping and easing or regulating at the front touch bushings should be carried out just as for the upright piano regulation (step 7).

Key levelling

Key levelling leads (lead weights with spring clips) can be attached to the back checks and the keys levelled by adding or removing balance papers under the balance washers. A number of test notes should be set-up first, to ensure the height of the keys is suitable when the keyboard is *in situ* in the piano, with the key slip in place. The height of the underside of the front of the key above the key bed will vary depending on the piano. Steinway models S - B are at 63 mm and the models C and D are at 65 mm,[1] whilst other instruments may typically be 71 mm. Whatever the measurement, it is aesthetically necessary for the bottom front edge of the key to be 2 - 3 mm below the top of the key slip. Levelling itself is carried out using a straight-edge, or an edge arched 1 mm higher in the middle.

Depth of touch

At this stage the depth of touch can be regulated to the default depth, usually 9.5 mm or 10 mm. The thickness of the touch baize must be such that some papers are employed, so that the touch depth can be slightly increased later on, if necessary.

Hammer travel

The action must now be refitted to the keyboard, and screwed down. The key keeper should be left off. The hammers must be tested for travel. The spacing between any two adjacent hammers at the hammer noses and hammer head tails, must remain the same as the two hammers travel towards the strike rail. If a hammer does not travel vertically, flange paper must be added underneath the hammer butt on the side towards which the hammer travels (viewed from the keyboard).

Hammer spacing

The action with keyboard should then be returned to the piano so that the hammers can be spaced to the strings. This regulation must be done in conjunction with a regulation of the *una corda* pedal, and in particular, the rest position of the action (usually to the left). Both the operation of the *una corda* and the hammer spacing when the *una corda* is not in operation, must be correct. When not in operation, each hammer must align centrally and symmetrically with its string course. When the *una corda* is in operation, the hammers must strike two of the three strings on the trichords, and one of the two bichord strings. When the *una corda* is not in operation, no hammer for a monochord string must be offset even

[1] Dietz, FR, *The regulation of the Steinway grand action*, Frankfurt, 1963

slightly to the same side as the shift. Check in particular that the hammers at the ends of sections, next to the frame bars, do not touch the frame at any point of motion.

Lever travel

With the action back on the regulating table, the levers must now be aligned with the rollers. The gate in the repetition lever must lie centrally under the roller. Observing from the keyboard side of the action, a vertical strip of flange paper must be placed under the lever flange on the opposite side to the gate's offset against the roller.

Jack position in the gate

Any jack that does not lie centrally from left to right in its repetition lever gate must be corrected. Lift up all the hammers to examine for this. If necessary, remove the hammer rest rail, and remove any such lever sub-assemblies with this problem. The bottom lever can be placed in a vice or secured against a solid surface, while the repetition lever is pushed down, exposing the top of the jack. The jack can then be struck carefully with a technician's hammer on its top, on the side towards which it needs to move, in order to locate it centrally in the gate. The lever sub-assembly can then be returned to the action, and the lever travel regulated if necessary.

The upper surface of each gate should be a paper's thickness above the jack top. Regulate this with the repetition lever regulating screw.

All the jacks should then be positioned so that the edge of the jack top furthest from the jack tail, aligns with the score mark across the gate. Viewed from the keyboard, the jack regulating screw is turned clockwise to move the jack top towards the key fronts, and anticlockwise to move it away from the key fronts. This regulation is a preliminary one, the fine adjustment of the jack position in the gate being made later.

Test notes

From this point onwards, it is advisable to carry out the whole of the remaining regulation on at least two test notes, before proceeding to regulate the whole action. When the repetition tests are reached, a revision of the checking, set-off and drop distances may be necessary if the action is not in perfect condition.

Blow distance

If the hammer rest rail was removed, it must now be returned, and in any case must now be set low. The hammers must not rest on the baize (or felt). The blow distance must now be regulated on the capstans.

Check first that none of the repetition springs are too weak to support the weight of the hammer, and that all the jacks are properly under their rollers. Any interference from the back checks must be dealt with. If weak repetition springs are found, it is wise to increase the tension in all beyond the normal degree. Tension the springs so that after checking, the hammer pops up positively. If this cannot be achieved, screw up the drop screw, and decrease the set-off distance by turning up the set-off dolly (both an anticlockwise turn viewed from above). Each hammer must rest always in the same rest position.

The default blow distance is 47.5 mm, which the distance from the *hammer nose*, to the underside of the strings, or strike bar on the regulating table. Vertical measurements taken in the action well, from the key bed to the underside of the strings, must be taken at the beginning and end of each stringing section, and transferred to regulating table. Be careful to ensure that the position of the action on the regulating table from left to right, places the hammers at the ends of each section exactly where the corresponding measurements for the height of the strike bar apply.

The hammer rest rail can then be set so that the surface of the baize or felt it is 3 mm below the hammers at rest.

Set-off

The set-off distances should now be regulated. Set-off is always tested on a *very slow motion rise of the hammer*, to find the *set-off distance*. It can be tested lifting the lever directly, rather then depressing the key, if necessary. Learners often do not test *slowly enough*. Set-off is where the hammer stop rising, even as the lever continues to rise, or the key continues to be depressed, and then falls back. It happens when the tail of the jack encounters the set-off dolly and the jack top is tripped back away from the roller. The *set-off distance* is the distance from the hammer nose to the underside of the strings (or the strike bar on the regulating table), at the hammers *closest* point to the string (or strike bar) before it falls back. The default distance is 2 mm in the treble, or half the thickness of the string, in the bass. Turn the set-off dolly down (clockwise viewed from above) to *increase* the set-off distance, or up, to decrease the set-off distance.

Problems and solutions in regulating set-off

Aftertouch should be present when the set-off is tested by depressing the key. *Aftertouch* is when the hammer drops back after set-off, using the key. If you cannot detect a set-off or *aftertouch*, you must find out why. First see if set-off and *aftertouch* can be found by lifting the lever directly. If it can, go ahead and regulate the set-off distance, lifting the lever rather than the key.

If there is still a problem, look at the jack tail when the lever is raised. Is it lower than the rest, with the jack tripped back? If not, you need to turn down the set-off dolly more.

If there is still no obvious set-off and *aftertouch* when lifting the lever directly, screw down the drop screw as far as it will go, and try again.

On some actions, there may still be no appropriate response, in which case weaken the repetition spring by turning the repetition spring regulating screw anticlockwise. If you weaken the spring too much, the hammer may not rise properly.

Now regulate the set-off distance, by lifting the lever directly, if necessary. Then turn the drop screw back up so that after set-off the hammer drops about 3 mm, leaving it about 6 mm from the string. On the slow motion set-ff test, the hammer should not go into check (with the hammer tail jammed against the back check).

Look at the drop screws. Are they approximately the same height? Will it be necessary to alter all of them before proceeding? This can be done quickly and approximately, if it is necessary, because fine drop regulation comes next. Look also at the repetition spring regulating screws. If you had to weaken the spring to allow set-off detection, it may be necessary to weaken them all, first.

Preliminary drop distance

Regulate the drop distances with the drop screws, on a slow motion set-off test, lifting the lever rather than depressing the key, if necessary. On the slow motion set-off test, after set-off, the hammer should drop to twice its set-off distance.

First checking

If there are no problems with checking, and the checking is already close to 16 mm, fine check regulation can be carried out, setting the checking distance to 16 mm (5/8th inch), omitting the fine checking regulation below. Otherwise, the purpose here is allow sufficiently good checking so that the repetition spring tension and drop distance can be regulated.

On medium strikes, the hammers should go into check 12 mm - 15 mm from the string (strike bar). The check distance must be the same for all hammers. Bending the back check towards the hammer reduces the checking distance. Be aware that there is a limit point in moving the check head towards the hammer, beyond which the hammer will fail to check, rather than check closer. This is because the hammer tail bounces off the top of the check head. Make sure the check heads are properly aligned left to right, with the hammer tails, and that they are not rotated. The check leather must be parallel to the hammer tail.

Problems and solutions

With medium strikes, the hammers should go into check. If the hammer fails to rise as the note is struck, screw the jack top towards the hammer, using the jack regulating screw (anticlockwise turn). If necessary, the repetition spring tension may be increased also, turning the regulating screw clockwise. The hammer may also fail to rise if the hammer tail encounters the back check as it rises. Bend back the check head if necessary.

If the hammer strikes the strike bar (string) but fails to go into check even on a hard strike, staying close to the string (strike bar), then bend back the check head a little. If the check head is too close to the hammer, the tail strikes the top of the check leather when the hammer returns from the string, rather than the hammer tail sliding down the fact of the check leather. The hammer may also fail to go into check if the check head is too far away from the hammer, in which case, bend the back check forward in small increments, until a checking occurs at about 2 cm or less, from the strike bar (string). If you try to reduce the checking distance too much, the hammer tail will just bounce off the top of the check head, rather then going into check.

First repetition spring tension

Provided all the hammers are already checking reasonably well, the repetition spring tensions can be regulated. On a medium strike, the hammer should go into check. On release of the key, it should rise positively, but not vigorously, up to the drop distance. As it rises, a sharp or hard knocking should not be felt in the key, but a soft or gentle sensation as the hammer rises, may be acceptable. All the hammers should rise in the same way.

Fine drop distance

The drop distances should now be fine regulated on the hammer rise after checking. Play a medium strike, so that the hammer checks. On partial release of the key (a small rise) the hammer should rise to the drop distance. Normally, this is the same as the drop distance already set on the slow motion test, but fine testing and regulating of the drop distance must now be done after the hammer rises, on release of checking.

Fine checking

The checking distances can now be fine regulated to 12 -16 mm (5/8th of an inch). The distance must be the same throughout. Closer checking will allow finer repetition regulation. However, on some actions, 16 mm may be the closest possible, for the correct depth of touch (up to 10mm), and whilst the depth of touch should allow *aftertouch*, it should not be increased in order to reduce the checking distance. Increasing the aftertouch will decrease the fineness of the repetition regulation.

Fine repetition regulation – *pianistic tests*

Fine regulation of the action must be carried out with the action in the piano, repeatedly withdrawing the action and replacing it, as necessary.

The jack positions under the rollers must now be fine regulated. This first requires the *pianistic* "misfiring" test (see Chapter 2 for an explanation of the pianistic tests). Strike the note abruptly and very hard. Screw the jack regulating screw clockwise in increments, continuing the "misfiring" test, until the hammer "misfires". On "misfiring" the jack shoots out from under the roller before the hammer rises, so no strike happens.

Now screw the jack back in, turning the screw anticlockwise, in small increments, until "misfiring" has just stopped. In this part of the test, each position of the jack must be tested with multiple strikes, to see if "misfiring" occurs intermittently.

Now the repetition spring tension must be re-tested, and adjusted, if necessary. However, now we must use the other opposing *pianistic* tests. The hammer must rise positively (not abruptly) after checking, but it must not be possible to induce hammer *bubbling*. Play without checking, using soft, tiny rises depressions of the key, as though reluctant to bed the key. If hammer bubbling can be *induced*, this requires a weakening of the repetition spring.

We must also try to induce *blocking* using the pianistic blocking tests. If it is possible to induce blocking through *secondary checking*, then the repetition spring tension should be increased.

On some actions, if a suitable repetition spring tension cannot be found that makes both bubbling and repetition blocks impossible to induce, then the other regulating distances may be altered slightly. Bubbling can be further avoided by increasing the drop distance. The set-off distance may then also need to be increased. Secondary checking blocking (repetition blocks) can be further avoided by decreasing the checking distance. If *aftertouch* is already present, this should not be done by increasing the depth of touch, but by bending the check wire. Remember that there will generally a *limit* to how small the checking distance can be made by bending the check wire, beyond which checking will fail.

If the repetition spring tension has been adjusted, the jack "misfiring" regulation may need to be carried out again. It may sometimes be necessary to "circulate" through the two regulations, repetition spring tension and jack position, to optimise the regulation.

Damper alignment

The dampers must be parallel to the strings, symmetrically placed over the string course, and perfectly vertical in two planes. The parallel alignment of the damper to its strings, can be fine regulated after the collective damper lift regulation. When the damper is dropped through its guide hole and positioned correctly on the strings, the bottom of the damper wire must be in line with the damper wire hole in the standard damper lift. If it is not, the wire must be cranked between the standard damper lift and the guide rail. Before cranking any wire, remove the damper from the guide rail, hold the side of the damper head opposite the wire against a vertical surface, and ensure that the wire runs parallel to the surface. Never bend the wire immediately where it exits the damper head. Bend only on the last bend before the wire drops away from the head.

Ensure also that the wire falls at right angles to the head looking at the side of the head. The damper can then be placed back in the guide rail, and cranking below the guide rail can commence. Cranking requires skill and patience. Compound leverage cranking pliers are the most effective tools for the job. Aim to crank the wire *before* inserting it into the standard damper lift. The length of wire that passes through the guide rail hole must not be kinked, and must remain vertical both planes. If necessary, lightly ream the damper guide rail hole.

Collective damper lift

The collective damper lift is the regulation of the dampers whey they are lifted using the pedal. On some actions an individual inverted capstan is provided on the underside of the damper lever, for this. Otherwise, the collective lift must be regulated by moving the standard damper lift up or down the wire. If there are no spoons, then make sure the chosen position of collective lift is suitable for the individual lift to occur when the hammer is half way to the strings.

The orientation of the damper parallel to its string course can be regulated by turning the damper wire. Use genuinely round nosed pliers for this, gripping the wire just above the standard damper lift.

Look from above, for twisting (clockwise or anticlockwise) or rolling (top of the head moves to the left or right) of the damper head, as it lifts or falls, and look for front-first or back-first lifting of the damper. If there is front-first or back-first lifting, or rolling, then the damper should be removed and the wire checked again for proper right angles in all three planes where it leaves the damper head. Twisting can be corrected using the round nosed pliers. Turn the wire in the same direction that the damper head twists as it falls.

There should be a small amount of lost motion between the damper lift rail baize or felt, and the bottoms of the damper levers or their capstans. All the dampers should lift simultaneously.

Individual damper lift

If the damper levers have spoons, then the individual damper lift can be regulated separately from the collective lift. If there are no spoons, then the individual lift must be combined with the collective lift regulation, and fine regulated by adding felt shims to the backs of the keys. The default individual damper lift should occur when the hammer is half way to the strings, but the point of damping as the key is released, must also be taken into account. The point of rise is not always the same as the point of completion of damping on return of the key to its rest position. Slightly inefficient damping may require a slightly later lift, and earlier return of the damper.

Una corda pedal

The *una corda* regulation should be carried out with the hammer spacing, and is outlined above. Additional adjustments may sometimes be necessary at the top of the pedal rod (lyre rod) or in the trapwork.

Damper lift pedal

The damper lift adjustment may be at the top of the pedal rod (lyre rod), and/or in the trapwork, and/or underneath the damper lift rail in the action well. The damper lift rail must not be acting on any damper with the pedal off. The lost motion may be varied according to the individual piano or pianist's requirements, but usually should be kept small. The pedal must be able to fully raise the dampers. The damper levers should not touch the damper lever stop rail when the pedal is fully depressed, or the key is fully depressed, slowly. The function of the rail is to stop increased damper rise on hard playing or use of the pedal.

Sostenuto pedal

Older *sostenuto* mechanisms have a fixed lip on the damper lift, whilst newer ones usually have the lip on an additional moveable lever. Innovatory variations exist. The essential function remains the same. When the sostenuto lever or damper lift is not lifted, rotation of the sostenuto bar places the bar's cam above the lip, so that it has no effect. When the damper is lifted first, then on operation of the pedal the cam encounters the lip on its underside, so that the damper remains lifted when the key is released.

If no other regulating function is available, the position of the lip can be regulated by bending the damper wire in or out immediately above the damper lever. The damper wire that passes through the damper guide rail must always remain vertical, so more than one bend may be necessary.

Additional adjustments may be necessary at the top of the pedal rods or in the trapwork.

7 – Very fine roller action regulating

The grand roller action can be very fine regulated for maximum sensitivity and regularity of touch in gentle playing, without compromising the available power in *fortissimo* playing. This necessarily requires first that the action is in good conditions, particularly the rollers, and that it is already regulated to a reasonable standard.

Assuming the notes are all working without bubbling or blocking faults, then the finest regulation begins with the principle of *minimum drop distance*.

The minimum drop distance is limited by the minimum available set-off distance, which must be smaller than the drop distance, and which is itself limited by the compression in the front touch baize and papers. Generally, a thicker baize, provided it is a hard one, is preferable to many papers, because it will be less compressible.

First, reduce the repetition spring tension until it is as weak as possible whilst still allowing hammer rising after check release, and without allowing secondary checking blocks (see the *pianistic tests*). Eliminating the possibility of secondary checking blocks, even through very "*unnatural*" playing technique to *induce* the blocking, is most important here.

Now regulate the jack position (jack regulating screw) until misfiring is possible, and then regulate it back until misfiring just becomes impossible.

Reduce the set-off until it is as close as possible, without allowing hammer blocking (hammer blocks against the strings) when the note is depressed in very slow motion, and after set-off, maximum weight is applied to the key front. In this test aim to squash the front touch baize and papers as much as possible with two fingers. Then play the note hard, and release the key so that the hammer is ready for a repeat strike. Now push up the hammer in slow motion with a very slow depression of the key, until it is as close as possible to the string (*i.e.* at the set-off point), to ensure the hammer does not touch the string when it is vibrating. If it does, the set-off distance must be increased a little.

Reducing the set-off will also reduce the aftertouch, so any depth of touch that is too small will tend to be revealed when the set-off is reduced. Any correction of the depth of touch may require the back check to be adjusted to maintain correct checking distance.

The depth of touch must not exceed its proper value (10mm maximum or 9mm on the Bluthner-type action, or 9.5mm on some Steinways models S - B). If the depth of touch exceeds its necessary proper value, this will work against fine repetition.

The feel of the set-off in the touch (the "notchy" frictional increase and release) can be reduced by reducing the difference between the drop distance and the set-off distance. Though not recommended, it *is* possible on some actions for the action to work with the set-off distance *larger* than the drop distance. When the set-off distance is equal to, or larger than the drop distance, the "feel" of the set-off in the touch is greatly reduced or eliminated, and this touch characteristic can indicate that this condition is present. Many pianists in any case *like* the feel of the set-off in the touch, so reducing it to vanishing point is not in general a good idea, though it can be done if required. Such a set-up can be very sensitive for allowing very fine gentle playing and repetition, but it must be remembered that the set-off itself actually needs to be as close as possible to the string for maximum sensitivity in soft playing. Usually, if the set-off distance is greater than the drop distance, then it is not as close as it could be. It is more advisable to start by reducing the set-off distance to a minimum as already described. The difference between the drop distance and the set-off distance can then be reduced as far as possible on the regulation of the drop distance, which we come to next.

Having regulated the set-off distance, the drop distance can be reduced as far as possible. The limits are:

1. The hammer must not bubble against, or touch the vibrating string when checking is released after a strong strike.

2. After check release, the hammer must not touch the vibrating string when the bedded key in pressed down with extra force.

3. The hammer must not bubble against the string on playing gentle, repeated notes, with only part of the touch depth.

Usually, the action will at this stage now be finely regulated and capable of the most sensitive responses. If, however, at any stage, secondary checking blocks (repetition blocks) become apparent, then the repetition spring tension may have to be increased. Any change to the repetition spring tension should be followed by a re-testing of the jack position (the misfiring test) and the drop distance.

If, at any stage, hammer blocking or bubbling becomes apparent without secondary checking blocks, then the set-off and drop distances may need to be re-tested.

If the touch depth is not too great, and all other regulations have been optimised, but sufficient sensitivity of repetition (the distance the key must rise before a repetition can be effected) cannot be attained, then the jack cushion in the top of the repetition lever gate should be examined to see if it needs to be thicker. Its function will be to limit the distance the jack can escape away from the roller, which will assist in repetition efficacy. In this respect it is similar in function to the jack slap rail of the upright action. If it is too thick, however, the jack will not set-off sufficiently, and hammer bubbling may result on soft blows.

From the above it can be seen that once the depth of touch and checking distance is correct, and assuming the jack cushion is correct, then the fine regulating of the action is made on the drop distance, the set-off distance, the repetition spring tension, and the jack position. Assuming the regulation is already fine, and the drop distance is greater than the set-off distance, then the following will apply:

Jack regulating screw

The distance the jack has to move back to get far enough under the roller to effect a repetition, needs to kept to a minimum. To achieve this the *aftertouch* needs to kept to a minimum necessary for set-off to be effective. This is why the depth of touch must not be too great. Similarly, the distance the jack must travel back, and the depth the key must travel through, from the beginning to the end of the set-off, must be kept to a minimum. This is achieved by setting the jack as far back on the roller as it can be positioned, without allowing misfiring.

Jack regulating screw

Clockwise turn	**Anticlockwise turn**
Tends towards:	
Misfiring.	Less sensitivity More pronounced feel of set-off in the touch.

Repetition spring regulating screw

The repetition spring tension must allow the hammer to rise after check release fast enough for effective repetition. If the tension is too great, the release of the repetition lever after check release is felt too much in the touch.

Repetition spring regulating screw

Clockwise turn	**Anticlockwise turn**
Tends towards:	
Knocking in the touch after check release.	Repetition blocks. Misfiring.

Drop screw

The drop screw limits how close the hammer can come to the string when supported by the repetition lever, the jack being set-off. It must be as close as possible to allow the most sensitive repetition. The hammer, however, must not come into contact with the strings except during an actual strike intended by the pianist.

Drop screw

Tends towards:

Clockwise turn	Anticlockwise turn
Decreased repetition sensitivity.	Increased repetition sensitivity.
Increased feel of set-off in the touch.	Decreased feel of set-off in the touch.
Repetition blocks.	Hammer blocking.
	Hammer bubbling.

Set-off screw

The set-off needs to be as close as possible, but if it is too close, it will allow hammer blocking on hard strikes, or hammer bubbling on soft strikes.

Set-off screw

Tends towards:

Clockwise turn	Anticlockwise turn
Decreased feel of the set-off in the touch.	Increased feel of the set-off in the touch.
Decreased repetition sensitivity.	Increased repetition sensitivity.
	Hammer bubbling.
	Hammer blocking.

8 – Steinway action differences

On the Steinway action the repetition lever regulating screw is at the bottom of the repetition lever, the button acting on the bottom lever. Clockwise or anticlockwise turning has the same effect as on the standard roller action, but the effect is generally more pronounced.

There is no bent block on the Steinway lever assembly, a spoon being in its place. The regulation of the jack is carried out in the usual way.

Steinway hammers have individual hammer rest felts rather than a hammer rest rail. As in the standard action, the hammers should be at rest above the rest felt, and not on it.

The Steinway action has set-off screws and buttons rather than dollies, but the regulation follows the same principles.

Perhaps the most important difference between the Steinway lever assembly and the standard one, is the repetition spring. There is no repetition spring regulating screw on the Steinway action, and the repetition spring is a "wing" spring that is supported by the balance block, its ends acting on the base of the jack, and the underside of the upper half of the repetition lever, where it rests in a slot. The spring should have its tension adjusted using a Steinway pattern repetition spring regulating tool, which is a hook device with a handle.

Increasing the tension:

Using the tool to press down on the upper part of the spring, the upper half of the spring can be released from its slot underneath the repetition lever, moving it sideways so that it clears the repetition lever to one side. The spring will tend to rise up beside the repetition lever. The hook on the end of the spring can then be pulled upwards to increase the bend in the spring and hence the tension when it is replaced. Bending the spring very slightly sideways towards the repetition lever will help to ensure the spring does not spontaneously dislodge itself from its slot after regulation. The technique requires some practise, and more patience than a regulating screw, but is just as effective when the right technique is used.

Decreasing the tension:

Decreasing the tension is a little easier. The hook on the tool is used to push downwards but not sideways, on the top half of the spring, pulling it out of its slot and decreasing the bend, so that when it springs back in its slot, the tension is decreased.

Remember that (in common with some other pianos) the removal of the fall, key blocks and key slip on the Steinway, can be made simultaneously, as they remain connected. To help prevent them falling apart when removed, the connection between the key slip and the key blocks can be tightened before removal.

9 – Regulating the Blüthner-type action

The depth of touch and blow distance in the Blüthner-type action are smaller than the normal default, being 9 mm for the depth of touch, and 45 mm for the blow distance.

The Blüthner-type action's repetition capability is as good as that of a well regulated roller action. The touch is generally "lighter", and "smoother", and the sensation of set-off in the touch is less pronounced.

Technique for handling the repetition spring

The repetition spring can be removed or replaced in its slot in the side of the abstract, using the index finger alone. The slot is angled at about 30 degrees, and the force from the finger should be applied on the spring in the appropriate direction, as close to the abstract as possible, *i.e.* with the finger tip touching the abstract. Once the technique is mastered, removing or replacing the spring is quick, and does not alter the regulated spring tension.

Key levelling

If the whole keyboard needs to be levelled, the action should be removed from the keyboard. If this is not necessary, levelling of individual keys can be carried out without necessarily removing the action from the keyboard on the Blüthner-type action, as the key can be lifted sufficiently high off the balance pin to allow access to it.

To separate the action from the keyboard, first disconnect the repetition springs by pulling then out of their slots in the abstracts, then remove the hammers together with their abstracts. Remove the rest of the action, leaving the keys, jacks, carriages and repetition springs in place.

The keys can then be levelled using key levelling weights clipped to the back checks. The key height should first be set on a number of "test" keys, with the action in the piano and the key slip in place. Where there is no vertical rise at the base of the sharp key coverings, the joints between the main part of the sharp key and its (ebony) covering should not be above the upper surface of the white keys.

Preliminary depth of touch

On the Blüthner action the default depth of touch is 9 mm. A preliminary depth of touch greater than this, *e.g.* 9.5-10 mm will assist with accurate set-off regulation later. Ensure that there are touch papers present under the front touch baize that can be removed if it becomes necessary to increase the depth of touch. The thickness of the touch baize should be chosen accordingly.

Fit the action to the keyboard

Return the action without the hammers to the keyboard and screw down the action standards. Make sure the lower nuts on the butt rail support bolts are in line with the upper surfaces of the action standards, then fit and tighten the upper nuts on the bolts.

Fit the hammers back onto the action, working from one end to the other, so that manipulation of the abstract can be obtained from access at the side. Be careful not to adversely bend the repetition spring when lowering the abstract, and keep it on the same side of the abstract as its slot. Lowering the hammer, insert the abstract pin into the hole in the back of the carriage. Be sure not to bend the pin. If the jack gets caught under the felt beneath the nose at any time, release it carefully by pressing down on the jack tail. Before screwing down the hammer butt, insert the repetition spring into its slot in the side of the abstract, in order to protect it from damage as the next hammer is fitted. Tighten the hammer butt screws if they are loose (over tightening can make subsequent hammer spacing more difficult). Test the tightness of the action standard screws and tighten any that are loose.

Hammer spacing

Adjust the *una corda* shift for shift distance. Space "test" hammers with the strings, at the beginning and end of each stringing section. Check that the abstracts of the "test" hammers are vertical. If not, alter the *una corda* rest position of the action and re-space the hammers to the strings, accordingly.

Hammer rest rail

In the Blüthner-type action the hammers will rest on the hammer rest rail. The rail and it cushion should be adjusted so the blow distance is 45 mm. If any hammers do not rest on the rail they can be lowered by adjusting the key carriage (see the next regulation, below) and if necessary removing the repetition spring from its slot in the abstract.

Carriages

Adjust each carriage (loosen the front screw, adjust the back screw, and then re-tighten the front screw) so that the jack top slips easily under the nose when the hammer is at rest. There should not, however, be any lost motion. Lost motion at the top of the jack on the Blüthner-type action will have a much more pronounced effect on the set-off than it does on other types of action, set-off typically being prevented by even a small amount of lost motion. This will result in hammer bubbling. Conversely, if the carriage is set too high, so that the jack top is not free to return under the nose, this can result in secondary checking blocks (see the *pianistic tests*) being easily induced.

Set-off

Remove the repetition spring from its slot in the abstract, leaving it at the side of the abstract. At this stage, aftertouch is required. If necessary, the depth of touch can be increased a little first, in order for the set-off point to be clear. The default set-off is 2 mm in the treble and 3 mm in the bass. Regulate the set-off using the normal slow motion set-off test.

Checking

On the Blüthner-type action no checking occurs on a soft strike. The checking distance for a medium-hard strike should be set to a default of 18 mm. Closer checking may be possible. A closer checking distance will help prevent repetition (secondary checking) blocks (see the *pianistic tests*) on repetition.

Remove aftertouch

Any aftertouch must now be removed by reducing the depth of touch until it just disappears.

Repetition spring tension

Replace the repetition spring in the abstract slot and regulate the tension so that after release of checking the hammer rises gently. The repetition spring tension can be increased by removing the spring from its slot in the abstract, and *slightly* increasing the angle between the vertical and horizontal parts of the spring. Fine decreases in tension can be made by pushing on the spring at the top of the vertical section, towards the abstract, using the finger tip. Even a *slightly* stiff centre in a hammer may mean the spring tension appears to be correct, when it is in fact, too strong. This will cause unpleasant knocking in the touch after check release (on jack return).

Fine checking

Fine regulate the checking distance if needed. The checking should be as close as possible to the string, and equal. The minimum checking distance is however subject to the normal limit point, beyond which checking ceases to be reliable, or fails, because the hammer tail encounters the top of the check head as it returns from the string.

Fine repetition regulation

Each note must now be tested using the bubbling, and repetition or secondary checking blocking, *pianistic tests.* Increasing the depth of touch and/or weakening the repetition spring will reduce hammer bubbling, but will increase secondary checking blocking. Decreasing the depth of touch and strengthening the repetition spring tension will reduce secondary checking blocking, but will increase hammer bubbling.

Hammer bubbling	*Secondary checking blocks*
Reduce the repetition spring tension provided the hammer still rises after checking release.	Increase the repetition spring tension, provided the hammer does only rises very gently after checking release.
Increase the depth of touch.	Decrease the depth of touch.

Once the repetition spring tension is correct in terms of the hammer rising *gently* after checking release, aim to finalise the fine regulation through the depth of touch.

If hammer bubbling occurs in the *pianistic test* for bubbling, do, however, first check the repetition spring tension. If it is clearly too strong (the hammer rises vigorously after checking release) then weaken it accordingly, and re-test for bubbling. Do not reduce the tension to the point where the hammer does not rise after checking release. If bubbling persists when the repetition spring tension is correct, increase the depth of touch.

Whenever a change to the repetition spring tension or the depth of touch is made, carry out *both* pianistic tests again to determine which way to continue making adjustments.

The final regulation should allow no bubbling and no repetition blocking.

10 - Regulating the D-type action

The D-type or *simplex* action is the easiest of the grand actions to regulate. It is a "spring and loop" action. The loop connected to the hammer butt passes through the jack, and the repetition spring which is connected to the jack, hooks onto it. The action is as much capable of sensitive repetition as the roller action, when it is well regulated. The repetition spring tension regulating screw (a grub screw) is in the bottom of the jack. The repetition spring should be tensioned so that even if the grub screw is removed, the tension is still able to support the hammer.

Key levelling

The action can be removed from the keyboard if complete key levelling and/or keybed cleaning is required. Levelling and depth of touch can then be carried out in the normal way (see the method for the roller action). The depth of touch can be set to the standard 9.5 mm.

Blow distance

If using a regulating table, the heights of the strings above the key bed must be transferred to the strike bar on the table in the usually way (see regulation of the roller action).

The blow distance can be set to the standard 47.5 mm. The blow distance is set, as for the roller action, by adjusting the capstans. The capstans on the D-type action may be a little more difficult to access, and a suitable tool can be made from a damper wire. The hammer rest rail should be set to approximately 2 mm below the hammer shanks.

Set-off

The set-off can be set to 2 mm in the treble and 3 mm in the bass. There should be *aftertouch*, and if this is not apparent, weaken the repetition spring if necessary (this is more likely in the higher treble), and/or increase the depth of touch.

Checking

The checking distance should be regulated to 12 - 15 mm.

Repetition spring tension

The repetition spring should now be adjusted so that the hammer rises after release of checking, but only as far as the set-off point. A clockwise turn of the screw *increases* the tension; anticlockwise *decreases* the tension. Test for the symptoms and necessary corrections in the table below.

Symptom	*Correction*
Hammer does not rise to the set-off distance.	Increase the spring tension.
Hammer rises and strikes the string.	Decrease the spring tension.
Hammer can be blocked against the string after release of checking, by pressing hard on the key.	Decrease the spring tension.

Pianistic tests

Test for secondary checking blocks (repetition blocks) which will be possible if the repetition spring is too weak. Increase the tension in small increments, if necessary, and test for the conditions in the table above, so as not to over-tension the spring.

Test for bubbling which will be possible if the spring is too strong, and/or if there is insufficient depth of touch. The depth of touch can be tested by looking for *aftertouch* in a slow-motion set-off test. If obvious aftertouch is present, then depth of touch is not the problem, and the spring should be weakened. If it is not possible to weaken the spring without allowing secondary checking (repetition) blocks to be induced, then the set-off distance may need to be increased until bubbling stops.

11 - Diagnosis

It is often the case that whilst faults in the regulation of an action may be preventing the action from performing as well at it could in its current condition, complete servicing or regulation of the action may not be an option at the time. One does not necessarily always want to have to implement an entire regulating "sequence", starting with a change of blow distance, in order to address shortcomings in the action's performance. Often, it is appropriate to approach the regulation more from the point of view of providing an immediate improvement in the action's performance, in the minimum available time. One of the things that distinguishes the approach of the experienced technician is the ability to *diagnose* faults and ascertain the measures necessary to correct them, or to improve the action's performance.

Diagnosis is in essentially two stages, or of two kinds. Firstly there is diagnosis that can be carried out without withdrawing the action from the piano, if it is a grand piano, or without removing the casework, if it is an upright piano. One can tell a surprising amount about the condition of an action just by playing the instrument. Secondly, further diagnosis is possible when the action itself is examined.

The quickest approach is to *listen* to the sounds produced, *feel* the touch, and then *look* at the action. In the case of visually impaired assessment *looking* at the action can mean tactile "looking".

Listening

The question is "What can be heard that sounds abnormal?" The commonest abnormal sounds caused by regulation failure are *hammer bubbling*, *hammer blocking*, *damper lift failure* (the damper does not lift sufficiently), and *damping failure* (the damper does not damp efficiently). One must be able to recognise these instantly from their sounds. Let's look at some examples.

There is a difference between the sound of the string being struck by the hammer but the hammer not bouncing off the string, and the sound of the strings being struck, but the damper not lifting. Both create a dull, damped tone, sometimes with a "buzzing", but each sound is distinctive and should be recognised when heard.

The *hammer block* may sometimes include some limited *hammer bubbling*. On an upright action, hammer bubbling will have amongst its likely causes, checking failure, partial set-off failure, and butt return spring failure. On a grand, it may be caused by partial set-off failure and/or an overly strong repetition spring. The set-off, checking and spring then needs to be examined.

On the upright action, hammer blocking will be caused possibly by checking that it too close, or a complete set-off failure. This is turn demands examination of the check head position, the depth of touch, and the set-off button regulation. On the grand, it could be caused by set-off failure or an overly string strong repetition spring, so it demands examination of the jack position in the gate, repetition spring tension, and set-off regulation.

If the sound indicates the damper is not lifting, rather than a hammer block, then clearly the damper regulation needs to be examined. With some experience, we can get confirmation from the *touch* whether the spoon requires regulating.

Touch

The damper that fails to lift sufficiently, or lifts too early, because of the spoon regulation, will be detectable in the touch. The touch has five distinct phases, which are:

1) The initial key depression,
2) The change in touch when the damper starts to lift,
3) The set-off, which on roller actions and many upright actions is felt in the touch;

and as the key rises:

4) The repositioning of action parts ready for repetition
5) The return of the damper to the strings

Phase (4) is usually only of consequence on the grand roller action. The rising of the hammer and the return of the jack top under the roller after release of checking can be felt in the touch, and if the repetition spring is too strong this is clearly identifiable in the touch.

The point of damper lift can significantly affect the perceived "weight" of touch, the touch being very noticeably "heavier" if the damper is lifting too early. There is a distinct "feel" in the touch when the damper lifts to early. Conversely, should the damper not be lifting at all, this can often be felt as a relative "lightness" in the touch when compared to other notes in which the damper is working properly, or when judged simply in the knowledge of the proper touch.

Some of the more common conditions for the upright action are listed in the following table. More than one symptom may be present, simultaneously. Remember that the *collective damper lift* should always be checked *before* the spoon regulation. Lift the dampers by the pedal or damper lift bar and make sure the damper wires are properly regulated first.

Also remember the following two techniques.

1) The set-off can always be tested by lifting the lever directly with the finger, and *should* be tested in this way if there is ever any

doubt as to whether set-off is taking place, or taking place early enough. This eliminates confusion with other causes such as insufficient depth of touch.

2) The key can isolated from the rest of the action by lifting the lever off the back of the key. Any sticking in the key can then be identified separately from other causes in the action.

Upright action

Touch	*Sound*	*Symptom*	*Condition to test*
Frictional resistance	Normal		Isolate key and test key bushings.
Frictional resistance / very "light" when note does not play.	Normal or note plays weakly	Repetition failure	Jack centre; Notch; Hammer centre; Isolate key and test key bushings.
Very "light"	No sound	Strike failure	Jack centre; spiral spring
Frictional resistance	Normal or soft "buzzing" on key release	Damping inefficient	Damper centre
Light	Normal	Damping failure	Damper spring; damper centre
"Heavy spongy"	Normal	Possible no damping	Damper lift (too early)
"Light"	Normal or damped	Possible permanent damping	Damper lift (too late)
Possible notchy set-off	Normal	Click when key rises after strike	Missing notch cushion
Possible notchy set-off	Normal		Notch leather; Jack top lubricant
Normal	Hammer	Bubbling	Blow distance;

	bubbling		depth of touch; set-off; checking; butt return spring; jack slap rail
Normal	Damped strike	Hammer blocking	Set-off; checking; depth of touch
Heavy (inertial)	Double note		Jack alignment (may be lifting two hammers); hammer spacing
Normal	Normal	Key sticks down	Isolate key and test key bushings; key sticks on lock rail when depressed; back of key touches adjacent key; swollen key leads; foreign body between keys
"Lost motion" feel	Normal		Lost motion

Older upright actions.

One of the commonest symptoms on older actions is hammer bubbling due to wear *etc.* It may not necessarily be appropriate to regulate the whole action starting by re-setting the blow distance. Where bubbling is present, it can be tackled by:

1) Testing the set-off distance by lifting the lever directly, and correcting it if it is too small.
2) Eliminating any lost motion
3) Either increase the depth of touch, and then correcting the checking distance, or:
4) Increasing the set-off distance.

Some of the more common conditions for the grand roller action are listed in the following table.

Grand roller action

Touch	*Sound*	*Symptom*	*Condition to test*
Frictional resistance	Normal		Isolate key and test key bushings.
Frictional resistance / very "light" when note does not play.	Normal or note plays weakly	Repetition failure	Jack centre; Hammer centre; Isolate key and test key bushings; Set-off; Jack position; Repetition spring tension
Very "light"	No sound	Strike failure	Jack centre; Hammer centre; Set-off; Jack position; Repetition spring tension
Frictional resistance	Normal or soft "buzzing" on key release	Damping inefficient	Damper lever centre; Damper guide rail bushing; Damper wire incorrectly bent
Light	Normal	Damping failure	Damper always up? Damper lever centre; Damper guide rail bushing; Damper wire incorrectly bent
"Heavy" as "weighty"	Normal	Possible no damping	Damper lift (too early)

"Light"	Normal or damped	Possible permanent damping	Damper lift (too late)
Set-off too notchy	Normal or difficult to play very soft	Possible noisy clunk on set-off	Roller worn; Jack position too far under roller
Resistance then sudden release	Normal or difficult to play softly		Check head catching hammer head as it rises
Normal	Hammer bubbling	Bubbling	Blow distance; depth of touch; set-off; drop distance; checking; repetition spring tension; jack position
Normal	Damped strike	Hammer blocking	Set-off; repetition spring tension too great; drop distance too small
Heavy (inertial)	Double note		Hammer spacing; Lever spacing - two roller being lifted
Normal	Normal	Key sticks down	Isolate key and test key bushings; key sticks on lock rail when depressed; back of key touches adjacent key; swollen key leads; foreign body between keys

12 – Brief notes on toning

Detailed instruction on toning would be beyond the scope of this book. However, some general awareness of it is appropriate for all technicians whether or not they would specialise in it.

Toning or voicing of the hammers is not something to be undertaken lightly, and it should always be approached as though it were an irreversible process. Ideally, it should be carried out by a technician who is also a master tuner. Only the latter has the "ear" and audiophile knowledge of piano tone necessary to achieve voicing to the highest standard, without loosing the potential of the instrument.

Before the felt is applied to the wood it has the form of a long strip. Its length matches the length of the moulded wood that will be cut across its width multiple times to form the hammer heads. The felt strip is initially triangular in cross-section, the size of the triangle graduating from one end of the strip to the other. The largest end will form the bass hammers and the smallest end, the treble hammers. The top apex of the triangle is aligned with the side of the wooden moulding that will provide the noses of the hammers, and the felt is then forced around the moulding, so that the base corners of the triangle fall towards the back of the wooden part of the hammer heads. This causes compression in the felt perpendicular to the nose, but extension forces around lines parallel to the curvature of the wooden nose. Subsequent insertion of needles into the felt in the toning or voicing process, releases some of this tension in the felt.

Worn hammer felts may need to be re-faced or replaced (or the hammer heads replaced) before regulating. Re-facing the heads by removing grooved and worn felt is achieved by use of a hammer file (an abrasive covered flat surfaced stick). Re-facing can only be carried out if there is sufficient depth of felt between the nose and the wood, and this is best checked in the upper treble, where that depth is smallest.

One of the problems created by grooved hammer felts is that the hammer will not always strike the strings of a trichord in precisely the same way each strike, and a slight change in alignment of the hammer can have a profound effect on tone, if the grooved do not perfectly align with the strings. These effects can make the finest control of tone through tuning, more difficult or impossible.

Hammer felts can be hardened, or can have their density increased, by heat-ironing (using a *hammer iron* tool designed for the purpose) or by applying a volatile fluid hammer hardener, or "dope", available from the specialist suppliers.

Softening of the felt, or controlled reduction of its internal tension, is achieved by inserting toning needles into the felt. The hammer is constructed by forcing the felt, triangular in cross-section, over the wooden hammer head, from a position in which the apex of the triangle points into the nose of the wooden hammer head. This creates high compression through the felt onto the nose, and places the fibres of the felt under tension pulling each side away from the nose and round the hammer.

The nose of the hammer is thus a critical structural area, and changes to the tension or physical properties in this area could affect the entire felt, and in an uncontrolled way. Doping or needling on the very nose of the hammer is therefore avoided.

Needling the felts has three aspects, which are (1) shallow toning, (2) medium toning, and (3) deep toning. Shallow toning changes the tone for soft playing, medium toning for medium strikes, and deep toning for loud playing.

Needles are inserted from the curved surface of the felt towards the base of the wooden nose (rather than the wooden tip). The part of the surface of the felt into which needles are inserted extends towards the nose, from just behind the level of the wooden nose. The very nose of the felt itself, as already said, is avoided.

Insertions furthest from the nose of the felt, are deep, and those closest to the nose are shallow, with the medium in between. Usually, deep toning is followed by medium toning, which is then followed by shallow toning. Toning is a *gradual* process involving many insertions of needles, and needles are frequently discarded and replaced as they become blunt.

The outcome of toning is to produce beauty and regularity of tone over the whole compass, for all levels of striking. It is a "circular" process to be carried out in conjunction with fine tuning. The finest tuning cannot achieve the optimum tone until the physical toning (needling) is optimised, but the needling process cannot be finalised unless the piano is in excellent tune. Just as tuning can change the tone, and even compensate to some extent for "hammer irregularities", so changes to the hammer felts will affect the finest tuning requirements. Generally, then, the piano need to be fine tuned, toned, fine tuned again, and toned again.

Some theoretical principles of toning

The hammer felt is graduated - it is hardest in the treble, and softest in the bass, and high quality hammer heads have throughout most of the compass, underfelt beneath the hammer felt. The best quality hammer felt is made from Merino sheep's wool, and high quality felt is generally denser than lower quality felt. The purpose of more expensive, higher quality felt, is primarily improved tone.

Two important physical parameters determining tone, are the *area* of hammer felt that contacts the strings, and the *duration* of contact with the strings. The area of contact is obviously affected by the hammer shape at the nose, and the duration of contact is also affected by the quality and density of the hammer felt.

A more pointed nose will produce a brighter, harder tone, whilst a flatter nose will produce a duller, more covered sound. Heavier felts generally produce a better tone, but both shape and hammer hardness or density needs to be appropriate for the strings in question, and string properties vary over the compass. Thus, the higher treble hammers are lighter, harder, and more pointed, because the strings are shorter, lighter, produce higher frequencies in their fundamentals, and have faster decay times. The bass hammers, in contrast, are softer, flatter nosed, and heavier. The strings here are more massive, longer, produce low frequencies, and have longer decay times.

13 - Strike line alteration

The tone in the treble part of the compass is critically affected by the precise position of the strike line. This can be altered as part of the regulation process. In the grand piano, it can be done as a final stage, but in the upright piano should be done at the beginning. In both cases, the piano should be well toned and tuned before making changes.

In the grand piano, to bring the strike line closer to the *capo d'astro*, shims or cloths can be added in the back of the action well to the runner blocks against which the key frame fits. To move the strike line away from the *capo d'astro* shims or cloths will have to be reduced in thickness, or the runner blocks altered.

In the modern upright piano the feet of the action standards typically rest in adjustable cups. The main alteration of the action height is made on the treble standard, and some adjustment may be necessary on the central standard. The capstans will generally have to be correspondingly altered.

If the tone is weak with too much hammer noise, the strike line may be too close to the top bridge or *capo d'astro*. If the tone is weak and lacks brightness and clarity, the strike line may be too far from the top bridge or *capo d'astro*. The situation should, however, be ascertained by first making quickly reversible, temporary changes to the action position, so see in practice how small changes in each direction affect the tone for both soft and loud playing.

Glossary of regulation distances

Blow distance (strike distance; hammer stroke) - The distance from the hammer nose to the strings when the hammer is at rest before a strike.

Checking distance - The distance from the hammer nose to the strings when the hammer has been checked, after a strike.

Collective damper lift - The simultaneous lifting of all the dampers as the damper lift pedal is operated.

Damper lift distance - The distance from the hammer nose to the strings at the point where the damper starts to lift.

Depth of touch (key dip) - The distance the front of the key travels from fully raised to fully depressed.

Drop distance (grand action only) - The closest distance achievable from the hammer nose to the strings, when the hammer is supported only by the repetition lever. The repetition spring must be sufficiently strong for the leather pad at the top of the repetition lever to be in contact with the drop screw.

Hammer travel - The travel path of adjacent hammers, which should remain parallel throughout travel, and strike perpendicular to the strings.

Individual damper lift - The point of lift of the individual damper.

Lost motion - Any distance through which the key depresses from rest, without lifting the hammer. It appears as a gap between the capstan and the heel of action (upright), or between the top of the jack and the notch.

Key height - The height of the white keys above the key bed, lock rail (upright) or key slip (grand), and the height of the top of the black keys above the upper surface of the white keys.

Key levelling - The relative height positions of the upper surfaces of the white and black keys, respectively.

Repetition spring tension - The strength or tension of the repetition spring adjusted by the repetition spring regulating screw on the standard grand roller action, or by bending the spring on other grand actions.

Set-off distance - On a *very slow motion test*, the minimum distance from the hammer nose to the strings, before the hammer drops back away from the strings. Set-off is the mechanism of the jack top being tripped away from the roller (grand) or notch (upright) so that in a normal strike the hammer moves only under it own momentum over the set-off distance, rather than being pushed by the jack.

Tie tape tension - On the upright action, the position of the tape end and bridle wire relative to the balance hammer. The tie tape is not actually under tension in normal use. The tie tapes are designed to keep the levers up, and to prevent the jacks getting caught beneath the notches, when the action is removed from the piano and replaced.

Some suggested reading

Dietz, FR, *The regulation of the Steinway grand action*, Frankfurt, 1963

Matthias, M, *Steinway service manual*, Frankfurt, 1990

Pffieffer, W, *The piano hammer: a detailed investigation into an important facet of piano manufacturing*, Frankfurt, 1978

Mason, MH, *Piano action handbook*, Washington (PTG), 1971 (2nd Ed)